Col. Rex Applegate & Micha[el]

BULLSEYES
Don't Shoot Back

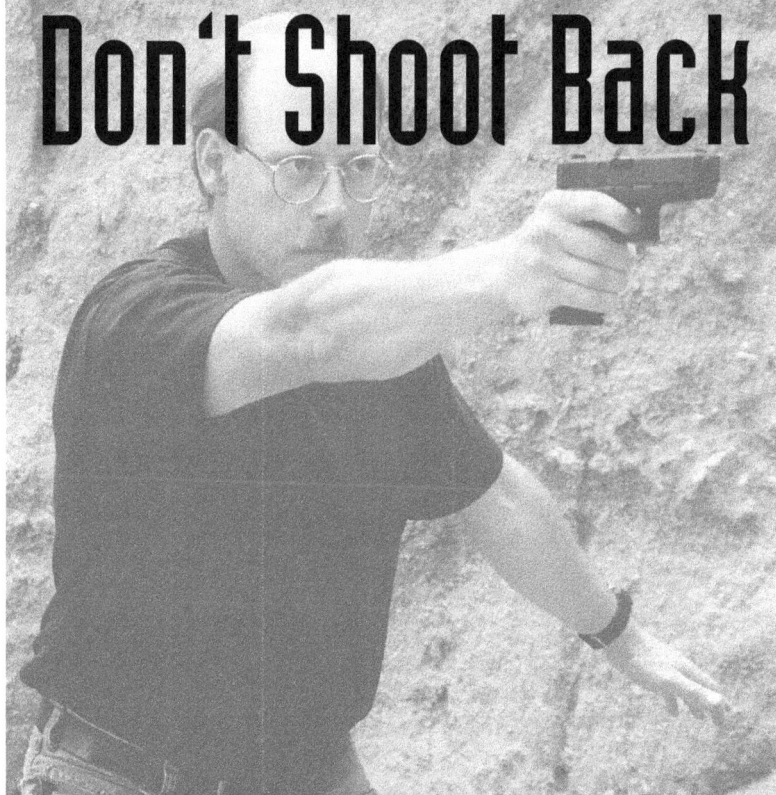

The Complete Textbook of Point Shooting for Close Quarters Combat

Other titles by Michael D. Janich:
Advanced Fighting Folders (video)
Blowguns: The Breath of Death
Breath of Death (video)
CBC: Advanced Concepts (video)
Contemporary Knife Targeting
Counter-Blade Concepts (video)
Counter-Gun Concepts (video)
Fighting Folders (video)
Focused Impact, Volumes 1-2 (video)
Forever Armed (video)
Homemade Martial Arts Training Equipment
Junkyard Aikido (video)
Knife Fighting: A Practical Course
Making It Stick (video)
Martial Blade Concepts, Volumes 1-6 (videos)
Mastering Fighting Folders (video)
Mastering the Balisong (video)
Mook Jong Construction Manual
Practical Unarmed Combatives, Volumes 1-3 (videos)
Silat Concepts
Street Steel
The Best Defense
Warrior Path

Bullseyes Don't Shoot Back: The Complete Textbook of Point Shooting for Close Quarters Combat
By Col. Rex Applegate and Michael D. Janich

Copyright ©1998 Col. Rex Applegate and Michael D. Janich
All rights reserved.

ISBN 978-1-939467-02-7
Printed in the United States of America

Published by Martial Blade Concepts LLC
1716A North Main Street
PMB 206
Longmont, CO 80501 USA

Visit our web sites www.martialbladeconcepts.com and www.martialbladeconcepts.tv.

CONTENTS

A Brief History of Point Shooting
  by Col. Rex Applegate xiii

Why Point Shooting? 1

Circumstances of the Typical Gunfight
and Practical Gunfighting Technique 3

Combat Shooting, Distance, and Time 15

Combat Shooting Phase 1:
  The Body Point 17

Combat Shooting Continuum Phase 2:
  Point Shooting 25

Combat Shooting Continuum Phase 3:
  Two-Handed Point Shooting 49

Combat Shooting Continuum Phase 4:
  Two-Handed Sighted Fire 53

Shooting from the Draw 59

Point Shooting vs. the State of the Art 63

Training Tips 65

Coming Full Circle:
  The Rediscovery of Point Shooting by Law Enforcement
  by Steve Barron, Hocking College, Ohio 73

Suggestions for Further Study 91

WARNING

F irearms are potentially dangerous and must be handled responsibly by individual trainees and experienced shooters alike. The technical information presented here on firearms handling, training, and shooting inevitably reflects the authors' beliefs and experience with particular firearms and training techniques under specific circumstances that the reader cannot duplicate exactly. Therefore, the information in this book is presented *for academic study only* and should be approached with great caution. This book is not intended to serve as a replacement for professional instruction under a qualified instructor.

PUBLISHER'S FOREWORD

This book began with the pre-production work for the Paladin video *Shooting for Keeps*. In 1994, Col. Rex Applegate, frustrated with the state of affairs of combat handgun training in the United States, proposed the production of a video on point shooting. Specifically, he wanted to produce a video that would educate the combat shooting community concerning the combat-proven method of point shooting he taught to the OSS and military intelligence operatives during World War II. In this video, he also wanted to put point shooting into proper perspective with so-called "modern" techniques that had been widely adopted as combat handgun doctrine.

Paladin's video production manager, Michael Janich, was tasked with organizing the production of this video. To ensure that Janich thoroughly understood the technique of point shooting and could faithfully present it on film, he received personal instruction in the technique from Colonel Applegate before beginning work on the project. Janich, a veteran Paladin author and avid shooter, then developed a narrative script for the video in close coordination with Colonel Applegate. This script formed the foundation for the most complete presentation on point shooting ever produced, as well as the most convincing case for the adoption of point shooting as a training doctrine.

Since its release in January 1996, *Shooting for Keeps* has been a best seller and has received widespread praise from firearms publications, law-enforcement agencies and the public. Bolstered by Colonel Applegate's tireless personal efforts to promote the technique, this video has also forced the shooting community to reexamine point shooting as a combat method. The result has been the formal adoption of point shooting by the police science department at Hocking College, a respected law-enforcement training facility in Ohio, and a renewed interest in the technique by other law enforcement agencies, civilians, and the military.

The heightened interest in point shooting has also prompted repeated requests for a written manual on the technique. To meet this need, Colonel Applegate and Janich collaborated to expand the *Shooting for Keeps* script into a book manuscript. In the process, they updated the text to include new developments, additional information, and a variety of supplemental training exercises. The result is this book.

Paladin has a proud tradition of offering diverse, sometimes controversial information to readers who seek to educate themselves and better prepare themselves to face the challenges of life. It is in the spirit of this tradition that we offer this book to you.

ACKNOWLEDGMENTS

The authors would like to express our sincere appreciation to Steve Barron and Clyde Beasley of Hocking College for their assistance and their written contributions to this work. We would also like to offer a special thanks to Bob Newman and Larry Hatem for taking the photographs included herein.

PREFACE

*B*ullseyes Don't Shoot Back covers the most practical, easily learned, and retainable technique for shooting the handgun in close-quarter life-threatening situations. This technique involves:

1) Facing the threat with both eyes open
2) Raising the weapon to eye level and firing instantly, with no front sight picture
3) Having a convulsive grip on the handgun
4) Firing mostly from a forward, aggressive crouch

When distance, light, cover, and time permit the use of sighted fire, the two-handed isosceles stance is recommended.

These techniques are directed primarily toward policemen, soldiers, and civilians who need to be trained in basic handgun skills for situations when the target is shooting back. The success of these techniques has been historically validated and documented.

The vast majority of military and police personnel simply are not interested in handgun shooting for recreational purposes, including target shooting, hunting, and "combat" competion programs. These soldiers and officers consider the handgun as just another tool, infrequently employed, during their professional careers. With few exceptions, the only things that most of this class of handgun shooter has in common with recreational-type shooters are training in handgun safety, maintenance of the handgun, and range procedures. This also applies to civilians who are solely interested in handgun training for self-defense purposes.

Over the past few decades, most civilians, law-enforcement officers, and military personnel have been (and are still being) trained in a system of handgun shooting described in the book *The Modern Technique of the Pistol* (Gunsite Press, 1991). Generally, this technique is based on the so-called Weaver stance. It advocates the use of two hands, an upright firing stance, separate trigger and breath control, and sighted shots for almost all close-quarter combat situations. It originally was designed to enable handgunners to achieve excellence in the sport of combat competition shooting, but it has little historical basis or validation for actual combat, where the realities of the battlefield and dark alleys involve individual shooter stress factors and instinctive reactions to actual combat situations.

To be declared adequately trained in the Modern Technique, the police or military recruit must spend hundreds of hours on the range and expend large quantities of ammunition to achieve "muscle memory" that purportedly will enable them to quickly, instinctively react in a real-life situation. Even after this goal has been declared to have been achieved, recent police experience shows that under combat stress, the shooter, no matter how well-trained in the Weaver stance, usually reverts to a crouch and one-handed shooting at close quarters and to the isoceles stance during situations that call for aimed fire.

Current available law-enforcement statistics also indicate that police—who have been trained almost exclusively in the Weaver stance technique for combat—have a hit average of approximately 15 percent. This can be attributed mainly to incorrect training techniques and, due to economic reasons, limited training time and ammunition expenditure on the range. Most police officers are now declared by their departments to be "trained" based solely on scores achieved against inanimate silhouette targets.

Conversely, programs that teach point shooting and the use of the isosceles stance for sighted fire will greatly reduce the financial costs of training time, ammunition expenditure, and the necessity for constant retraining programs now required by most departments. Officers trained in point shooting will also become more skillful, confident, and aggressive in firefight situations.

Recent scientific literature and on-going studies verify that only the most basic motor skills come into play during life-threatening situations. Other studies indicate that Weaver-type training involving multiple fine motor skills will not stand up in almost all actual close-quarter combat situations involving a handgun.

Police trainers and administrators are well-advised to approach this subject—so vital to the morale of all officers—with an open mind. They should not let the gun press, shooting "gurus," movies, television, or private for-profit shooting schools be the dominating influence when developing their own training programs.

A BRIEF HISTORY OF POINT SHOOTING
by Col. Rex Applegate

I started my training in combat handgun shooting under a couple of gentlemen named William E. Fairbairn and E.A. Sykes. These two men were the English police officers who operated in Shanghai between 1900 and 1940, when the city was one of the most lawless in the world. While serving in Shanghai, Fairbairn and Sykes developed the first comprehensive combat-oriented training for police use of the handgun, and they were co-authors of the first text on the subject, *Shooting to Live*.

Fairbairn rose from the position of constable to assistant commissioner, the second highest rank in the Shanghai Municipal Police Force. He not only had extensive command and training experience, he also participated in over 200 gun battles during his career. He combined this extensive real-world experience with an intense curiosity about the violent side of police work—why some things worked and others did not. He came to the United States in the 1920s and 1930s and visited the major police departments that were declared to have the most advanced U.S. police training for handling the violent side of police work, including those in Chicago and New York. Through his experiences and studies, Fairbairn developed many tactics and techniques in the early twentieth century that are standard practice today.

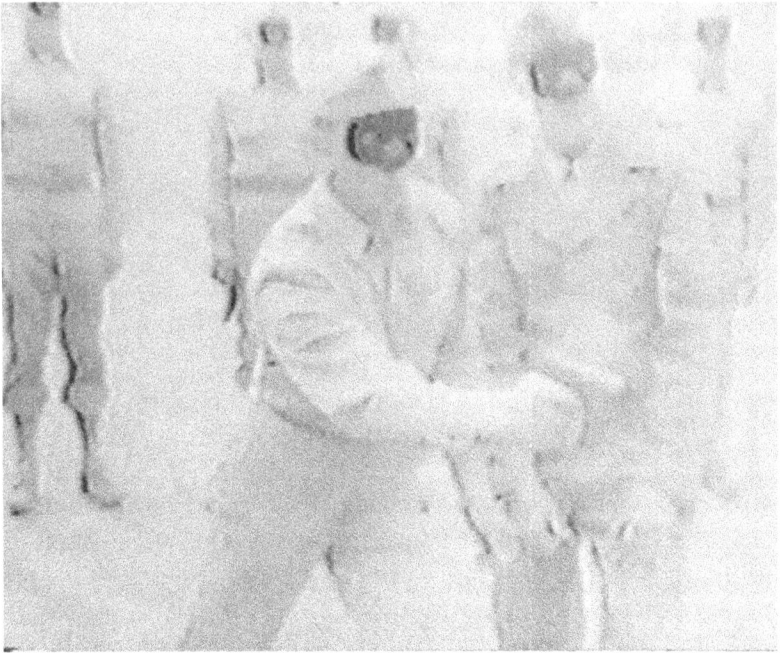

This rare photo taken from a declassified World War II training film shows British close-combat legend W.E. Fairbairn instructing an OSS agent-in-training in the finer points of point shooting with the Colt M1911.

TRAINING THE BRITISH HOME GUARD

In 1940, shortly after the outbreak of World War II, the British Army was ill-prepared and ill-equipped to handle the German blitzkrieg type of warfare. In France, the Germans pushed the British to the French coast of the English Channel at Dunkirk. The British executed a rescue mission using all available boats, but its army was forced to leave most of its armament on the beaches.

The problem was immense. Hitler was on the coast of France making plans for an invasion of England, and the British people were without weapons.

Fairbairn's original point shooting method only raised the handgun to the level of the chest. This method, though still effective at close range, is not as accurate as raising the gun fully to eye level.

This was when Fairbairn and Sykes were called in from Shanghai to offer their services and expertise based on the violence they had experienced in the Far East. Fairbairn had never put on a military uniform, and Sykes had served only briefly as a lieutenant during World War I. However, the minute they stepped off the boat, both were made captains in the British Army. Their first assignment was to train the British home guard (civil defense units) against what was perceived to be an imminent German invasion.

Due to stringent gun control laws, the British police and public were almost entirely without weapons. This was a country with very few licensed sporting arms, and so there was no immediate source of guns. Since the army had left the bulk of its weaponry on the shores of France, the entire nation was close to helpless. Fairbairn and Sykes had little with which to train the home guard.

The British law enforcement establishment, including Scotland Yard and the other government bodies, didn't believe in letting the public or police possess firearms. The English bobby was famous for, and bragged about, not carrying a firearm. The attitude was, "There's no crime in England; we don't need firearms, old boy." Fairbairn and Sykes were considered pariahs because they were so interested in violence.

Although elements in British law enforcement considered Fairbairn and Sykes to be barbarians for killing all of those criminals in Shanghai, the British War Office welcomed them. To help stand off an imminent German invasion, Fairbairn and Sykes taught the public how to fight with scythes, pitch forks, and improvised weapons as well as conduct guerrilla operations in the streets and countryside.

WITH THE COMMANDOS AND OSS

After about a year, Fairbairn and Sykes were assigned to train the British Commandos in armed and unarmed combat. Then in 1942, both officers were ordered to train British intelligence personnel operating in the underground in France and other German-occupied territories in Europe. From this point on, their activities were classified "top secret" and still are.

Later that year, Fairbairn was sent to the United States to help organize the combat training of the fledgling Office of Strategic Services (OSS). It was then that I first met him. He was, in my eyes, a crusty old bastard, about 57 years old, which I thought made him a hell of an old man. He was about 5 foot 8 inches in height and weighed about 157 pounds.

My first introduction to Fairbairn gives you an idea of

To give OSS students an idea of the stress they would experience by being on the receiving end of gunfire, Fairbairn would fire live rounds within a few feet of them during their training. Note the pistol in full recoil and the ejected cartridge case above the gun.

how ignorant some of us can be at times. I had just met him about an hour before he spoke at what is now the presidential retreat at Camp David, where the OSS training headquarters were located. I had been ordered by Col. William "Wild Bill" Donovan, the first director of the OSS, to be Fairbairn's assistant and learn all there was to know from him, pick his brains, and avail myself of his experience, then add to it and train our own people.

In one area of the camp was a meeting room with a small stage. A bunch of wooden folding chairs were set out in front

Colonel Applegate's association with W.E. Fairbairn was highly educational, but not always enjoyable. Here Fairbairn gives Applegate a lesson in the fine art of inflicting pain during the filming of an OSS training film.

of the stage for the hierarchy of the OSS (and future chiefs of the CIA), who had come down from Washington for a briefing. Fairbairn began by describing how he had done things in Shanghai and how they were now training in war-time Britain—learning how to assassinate people, blow up bridges behind the lines, and all the good stuff that happens in a total war. This was all top secret stuff and new to the Americans.

When Fairbairn got into the unarmed combat discussion, he said, "Lieutenant Applegate, come out here." I walked out on stage. He said, "Lieutenant Applegate, I want you to attack me." I said, "You can't possibly mean that, sir." He said, "I want you to attack me *for real*. That's an order." Like many second lieutenants, I thought I knew it all. I said to myself, I'll take care of this dumb old bastard. I let out a roar and went at him with both arms wide open.

The next thing I knew, I was flying through the air and headed for a landing on top of the members of the audience in the folding chairs. That made me very humble very suddenly and, those on whom I landed, very irritated. It was an introduction that has always stuck in my mind, because at that point, I decided I would listen rather than rely on preconceived opinions that weren't backed up by experience. I found, subsequently, that I usually learned something when I listened.

EARLY RESEARCH INTO POINT SHOOTING

With that introduction to practical, effective combat methods, I undertook the general assignment from Donovan to learn all that there was to know about armed and unarmed close combat. That was a hell of a big order! I'm still learning today.

If there was someone who was an expert at strangling people, or some guy who had killed a lot of people with a knife, I went to see him, even if he was in Sing Sing or San Quentin or wherever the hell it was. I had a unique opportunity to be exposed to the violent side of personal combat at U.S. government expense. Which brings me to the subject of handguns in combat.

My earliest introduction to handgun shooting came from Gus Peret, an exhibition shooter for Remington Peters, who was married to my aunt. Peret would come home to Yoncalla, Oregon, and do his practice shooting, much of which was of the old-time exhibition type such as practiced by Annie Oakley, Ad and Plinky Topperwein, Herb Parsons, and the shooters in the Buffalo Bill circus. I was the kid who threw the bricks in the air for Peret to shoot while he performed point or instinctive shooting.

I did a lot of plinking and handgun hunting as a young boy, but I wasn't introduced to any formal target shooting and training until I went to the University of Oregon and enrolled in the U.S. Army Reserve Officer Training Program.

I entered the army in 1939 as a second lieutenant of the Military Police Corps at Ft. Lewis, Washington. By the time I had my first lieutenant bars, I had been given a regular army commission. I had become what the army declared to be "thoroughly trained" with the U.S. Army Model 1911 handgun. Although I might not have thought about it very much, I remember having a feeling that it seemed like a pretty incomplete method to prepare a guy for battle. So when I got the assignment from Donovan, this was in the back of my mind.

I had access to most pre-World War II books, such as those by J.H. Fitzgerald and Ed McGivern. However, there were few written words on how to actually shoot handguns as the old frontier gunmen were supposed to have done it. There were plenty of stories about what happened in their gunfights, but there were never any details on the techniques they used when they shot the handgun.

The first indication of how the old timers did it came from a book I found called *Triggernometry* by western story writer Eugene Cunningham. It was a limited edition from a small publisher in Caldwell, Idaho. I went through this old book several times, and it gave me a few clues as to how the gunmen of the Old West shot, but most of it was assumption. Remember also that this was the time of the old cowboy movie stars such as William S. Hart, Tom Mix, and Hop-a-Long Cassidy. These men depicted shooting from the hip—killing Indians by "snapping" the revolver and knocking them off their galloping horses. This kind of stuff led to a public conception—mine and everyone else's—that this was the way to shoot at close quarters.

But going on information I had gleaned from *Triggernometry*, I went to Deadwood, South Dakota, where

"Wild Bill" Hickok was killed and buried. At the time I visited, it was a very remote spot, with none of the hype now associated with Wild Bill.

Hickok was an authentic Western gunman with a documented record of killing a number of men in face-to-face gunfights in the Old West. But by the time he was murdered, he was also somewhat of a national figure made famous by an author of dime novels, Ned Buntline. There are many stories about the mythical exploits of Wild Bill Hickok, including how he killed 18 Indians in a telephone booth and other fantasies. He performed on the vaudeville stage and in the Buffalo Bill Wild West circus but, although his career had been vastly overinflated by the dime novel press, he wasn't a phony.

I arrived in Deadwood in early 1942 and went to the old county courthouse. There was a little old lady there, and I asked, "Ma'am, do you have anything—newspaper articles, clippings, or any other information—on your most famous local character, Wild Bill Hickok?" She pondered awhile and said, "Lieutenant, I think we have something down in the basement." After about 20 minutes she came back with a dusty bundle of papers with a big old red ribbon tied around it.

I had a few hours before the next train, so I started going through the papers. Most of them were just old newspaper stories and repeats of the dime novel episodes. But deep in the stack were two letters fastened together by a straight pin. One was a letter from a character asking Hickok, in effect, "How did you kill those men? What was your method or technique?" This was exactly what I was looking for.

For some reason, Hickok's reply, in his own handwriting, was fastened to this letter of inquiry. Obviously, it had never been mailed. Hickok's response was, "I raised my hand to eye level, like pointing a finger, and fired." This was contrary to all the Old West shoot-from-the-hip techniques popularized on film and in the press.

This statement was very intriguing, but its significance was not made clear to me until Fairbairn and Sykes came along. With them came the experience from Shanghai as well as the experience of the British commandos and Special Operations Executive (SOE) branch of English intelligence on the use of handguns in combat. When I thought back to that letter, it was like I had discovered the Holy Grail.

CAMP RITCHIE

From my initial association with Fairbairn, I was ordered to England, where I worked with Sykes and others in operations before returning to the United States in late 1942. Shortly thereafter I was ordered to report to the U.S. Army Intelligence Training Center at Camp Ritchie, Maryland, where I organized the combat section for training in the same subjects I had specialized in at the OSS.

We trained 10,000 men in point shooting at Camp Ritchie. They were from all ethnic backgrounds, of all statures, with all sizes of hands, and with many different motivations. Many were brought to the training center not so much for their ability to fight but because of their language background or knowledge of German-occupied areas (this was also true in the OSS). We had these people coming through, and many times we were only given a few hours to train them with handguns—but we were successful. The Fairbairn/OSS/Camp Ritchie technique worked.

I had 28 officers and an equal number of noncoms working with me. We kept refining what the British had been doing while retaining the same basic principles. I sent not only trainees into combat but instructors, who would go out on special assignments and then come back to tell us if our techniques in shooting, knife use, strangulation, unarmed fighting, or whatever worked. If it didn't work, we threw it out.

We had school graduates going out into military units on assignments and then coming back. We had battle reports from our own intelligence and that of our enemies and allies. So, contrary to what some of the so-called "gurus" of today are stating, point shooting is not an untested, untried theory.

THE PROBLEM WITH SIGHTED-FIRE TRAINING

Several years ago, I was at the law enforcement trade show TREXPO. A young man got up on the stage to lecture on "Advanced Handgun Shooting." He said you put your foot forward here, and put your other foot back here, bend your elbow here, put your other hand here, control your trigger finger this way, and use the sights for all types of close-quarter shooting.

At the end of the class a SWAT team member got up and said, "Young man, why do you think this is the best way to shoot at close quarters in a gunfight or for training a policeman to shoot in the street?" The instructor said, "Because 50 of the world's greatest combat competition shooters shoot this way."

That's when I took notice, and it really got to me. The speaker certainly was not aware of the type of shooting the average cop does on the street and in the dark alleys. I remember talking to Michael Nossaman and some of the other people in attendance about what a shock it was to me that instruction in the Weaver style of sighted shooting had been going on for so long *to the exclusion of other methods*, and that there was almost no police or military training being given that did not involve the use of sights.

This is where I may differ in how to train police and military recruits for handgun combat. Possibly it is because my background is different. It's also true that I'm almost 82 years old, and my thoughts are not "up to date." But I have nothing to gain. Some years ago, I sent my son to Jeff Cooper's school. I knew what was going on there, but then I considered combat-style competition shooting, entailing hundreds of hours of practice and thousands of rounds of ammo to achieve excellence, as a sport. It was certainly not the exclusive way to train cops for duty on the street.

Currently, combat competition type of instruction is being given all over the United States by the military and local, state, and federal law-enforcement instructors. Yet reports show that most trainees can't hit their own ass under stress in actual close-quarter firefight situations.

The best records we have are the statistics from the New York City Police Department. Every year for the past 10 years, NYPD has gathered statistics on every firefight its officers have engaged in with "perpetrators," as they call them. The statistics are very complete as to the details of these gunfights—the time of day or night; whether or not the officer was kneeling, sitting, standing, aiming, or using one or two hands; what the light conditions were; what the ranges were; number of shots fired by the officer and perpetrator—the whole thing. The records also summarize each set of statistics with a brief description of the event. You couldn't ask for a better set of data from which to work.

Every year, the New York police average of "hits in firefights" is somewhere between 14 and 16 percent. That's pretty sad. Now maybe I'm being wrong or illogical. I know there are a lot of problems with police department administrations, and that many departments get locked into a handgun training program that only the chief wants, or the politicians require, or limited finances permit. But you would think that even in New York City, where there was a system of training in place that was failing over a 10-year period, after the first two or three years the people responsible would say, "Gosh, my guys are hitting only 14 to 15 percent of the time according to our own records. We must be doing something wrong. Let's try to improve on it." Well, nothing has happened, and it goes on and on. It was 14 percent again this last year.

I have come to the conclusion that either the trainers don't know what to do to improve the hit factor, or maybe they are restrained from doing it by their administrations. I have also come to the conclusion that many police officers *do not read*.

In 1943 I wrote a book called *Kill or Get Killed*. By a rough estimate, 250,000 copies are in circulation. It's now in its sixth edition, twenty-eighth printing. It has been made into a manual and is in current use by the U.S. Marine Corps. I wrote in there most of the things I am telling you now. Yet despite its widespread availability, the lessons and techniques in that book have not seemed to reach many law-enforcement trainers.

POINT SHOOTING TRAINING FILMS

In 1944, the U.S. War Department authorized an official training film entitled *Combat Use of the Handgun*. The black and white 16mm film (Film Bulletin 152) was professionally made by a Hollywood crew that had been sent to me at Camp Ritchie. It became an officially authorized training film for the U.S. Army and was classified "restricted." As late as 1951, it could be checked out from the army film library. That's the last time I saw it.

I went off to Mexico and a few other places and did not think much about the film for the next 40 years. However, after my observations at TREXPO, I thought I would try to locate the film. The army libraries didn't have it, so I went to the National Archives. There was a listing of it there, but no actual film. Six months later, I finally found a copy at Norton Air Force Base in California. I don't know how, or why, it ever got there.

After my discovery, the army had it declassified and sent me a copy. It is now available as a video from Paladin Press called *Point Shooting*. I did an introduction covering background information and, more or less, the things I am saying here.

The current U.S. Army trainers apparently know nothing about it. It certainly did not take too long for this film and the technique it teaches to disappear completely from military training circles. Last year I decided to update the video, and in January 1996 Paladin Press released another video called *Shooting for Keeps*. It's about an hour long and brings point shooting up to date, as I see it, for civilian, police, and military instructors. I would also like to recommend that you purchase a book by Bruce Siddle, *Sharpening the Warrior's Edge*, which also validates the effectiveness of point shooting.

POINT SHOOTING TODAY

What do we have in the form of police and military recruits today? We have people of all physical shapes, with different hand sizes, from various ethnic groups, and of both sexes. The majority of them have no special interest in handguns. They are not hunters, target shooters, or combat competition participants. To them a handgun is just another tool hanging on their belt. The more urbanized we get, the more this is true. It is particularly true east of the Mississippi River (sheriff's departments in the south and west still have a few recruits with shooting and hunting backgrounds). So this is what you have to work with—recruits who don't care about and have no knowledge of handguns and firearms. As an instructor you're going to have to do the best you can to interest them in correct training as a matter of personal survival.

One thing I don't particularly care for, although I know it is politically correct, is the training term "defensive shooting." The inference is that you wait for the other guy to shoot, then you shoot back or otherwise take action. I think police courses should be called "Practical Combat Shooting."

Statistics indicate that 60–80 percent of all police handgun firefights with criminals take place at distances of 20 feet or less, under conditions in which there is little or no light to see the sights, no time to see the sights, or no opportunity to use the sights. Irrespective of this fact, almost all civil and federal law enforcement agencies as well as the military do not train in unsighted shooting techniques. Accordingly, shooters miss at close range because of faulty training. Training almost exclusively in shooting at bull's-eyes and silhouettes using a two-handed Weaver stance *has got to stop!* You owe it to the recruits, and you owe it to yourselves as trainers. *Train to counter what your statistics tell you is happening in real life.*

Let's see if we can get this thing turned around. Let's initiate a program that includes training time devoted to the isosceles stance *and* point shooting. Everybody knows that you can shoot better with two hands if you use the sights and have time to aim than you can using one hand without the sights. But we must consider the cop (or whoever) in a dark alley whose adrenaline is pumping and who is facing a sudden muzzle flash along with a loud "bang."

We should also not have recruits trying to achieve good qualification scores on bull's-eye or silhouette targets. I know that is how police combat abilities are graded many times. I also know that "good scores" are often what the police department defense attorney wants in court when he defends a liability suit. ("This man is trained. He shot 90 percent on official silhouette targets.")

If I were an instructor and I had a class of recruits that was a mixture of males and females, of different physical sizes and ethnic backgrounds, and from urban areas with no firearms background, I would start the course with required safety procedures, then progress directly to close-quarters point shooting. Disinterested students can relate better to their own well-being and maybe won't go to sleep in class. Finally, I would only teach two-handed sighted shooting using the *isosceles stance* because, regardless of how long you train in the Weaver stance, statistics show that shooters revert to isosceles under life-threatening situations.

Point shooting is simple for students to comprehend and learn. It basically consists of a convulsive grip on the firearm, a rigid wrist, a rigid elbow, and a shoulder pivot point. You raise your gun hand to eye level and go "bang." No separate trigger finger action, no recoil problems. It's like pointing your finger.

At this time, an extensive test program on point shooting is taking place at Hocking College in Nelsonville, Ohio. Initial reports indicate that it is an outstanding success. The police science division of the college is training police recruits and students in this discipline and retraining officers who had previously been trained in Weaver only. All are being required to meet the Ohio State Police standards. Each type of student is being compared against the other. I might also add that, for training emphasis, they are using handguns on which the front sights *have been filed off or taped over.*

Proper training in combat point shooting achieves quicker expertise, does not necessitate so much retraining to maintain proficiency, saves more police lives, and takes more criminals, permanently, off the streets.

Then Lt. Col. Rex Applegate at his desk, Section VIII, Camp Ritchie, with the tools of this trade—silenced pistols, U.S. and enemy grenades, truncheons, and ammunition.

WHY POINT SHOOTING?

The term "combat shooting" creates very different images in different people's minds. To some, it is nothing more than shooting for scores using a human silhouette target rather than a bull's-eye. Firing from known distances often as far as 25 yards from the target, these shooters strive for tight shot groups and X-ring hits.

Others believe combat shooting is best represented by IPSC-style matches where shooters compete on contrived fire and maneuver courses, often using handguns that are so heavily customized that they would be impossible to carry for personal defense. In this style of shooting, speed and accuracy are valued above practicality and realism, and shooters are practically forced to abandon sound gunfighting tactics if they want to earn a respectable score.

A few shooters do take a wholly practical view of combat shooting and gear their training entirely toward realistic gunfight conditions. They train to shoot at unknown ranges from a variety of shooting positions and sometimes even shoot in subdued light. However, in almost all circumstances, they emphasize the use of carefully aimed fire accomplished by concentrated focus on the weapon's sights.

The problem with these interpretations of combat shooting is that none of them bother to consider what actually happens to a shooter *when the target is shooting back.* There is a tremendous difference between shooting methods that work well when you're simply trying to put holes in the target and those that work well when the target is trying to put holes in *you*. Failing to understand this difference is a mistake that will get you killed if you ever have to use your handgun in a real armed confrontation.

CIRCUMSTANCES OF THE TYPICAL GUNFIGHT AND PRACTICAL GUNFIGHTING TECHNIQUE

The handgun was conceived and designed as a close-range weapon. The history of its use as an offensive and defensive arm by private citizens, lawmen, and members of the military has repeatedly confirmed its effectiveness at this range and established it as the dominant close-quarter combat weapon. This history has also shown a clear pattern with regard to the circumstances in which a handgun will typically be used. The vast majority of recorded handgun shooting incidents have had three things in common: they occurred at very close range, in low light, and under conditions of extreme stress.

Based on these historical consistencies and the fact that the nature of gunfights has not changed significantly since the inception of the handgun, it makes sense that most modern confrontations involving the handgun will also occur under these basic circumstances. Recent statistics confirm this.

The FBI's 1992 *Law Enforcement Officers Killed and Assaulted* report provides details of officers feloniously killed by firearms over a 10-year period. According to this report, during the period considered 500 officers were killed by handguns, 94 by rifles, and 56 by shotguns. Of the 650 officers killed, 367 were shot at ranges of 5 feet or less, 127 at ranges of 6–10 feet, 77 at 11–20 feet, and 79 at 20 feet or more.

If we look at percentages based on these statistics, if you are involved in a gunfight, *there is an 88 percent chance that it will occur at a range of 20 feet or less.*

The "combat crouch" incorporates all the body's instinctive reactions to extreme stress, including crouching to present a smaller target, squaring your body with the threat, focusing your vision intently on the threat, and convulsive muscle contractions. Since it represents a natural reaction to stress, it is the ideal platform for all combat shooting technique.

The FBI study also revealed that 62 percent of the officers killed were shot between 6:00 PM and 6:00 AM, during the hours of darkness. Based on this fact, we can reasonably assume that the typical gunfight will occur in low light conditions.

Needless to say, if you are using a handgun in personal defense, you will be doing so because *you fear for your life.* Fear of this type causes extreme emotional stress and a series of predictable and unavoidable physical responses. These responses are instinctive gross motor movements that will almost always override your ability to perform finely coordinated actions. These instinctive reactions include:

• crouching to present a smaller target

- squaring your body with the threat
- focusing your vision intently on the threat
- convulsive muscle contractions

The only way to overcome these instinctive reactions in a high-stress situation so you can utilize finely coordinated shooting techniques is through continuous, *intensive* training. Such training must be challenging enough to generate levels of stress that approach those experienced in an actual gunfight. If these high levels of stress are combined with repeated drilling of combat shooting technique, the body can develop the ability to perform that technique under stress. The problem with this approach is that it requires countless hours

of initial training, continuous reinforcement training, and thousands of rounds of ammunition. While this may be feasible for competition shooters, SWAT team members, and elite military personnel, for the average citizen or law-enforcement officer with limited time and resources for training, it is not practical.

Moreover, no artificially induced stress factor can duplicate the fear and tension of a real life-threatening encounter. It is therefore entirely possible for even highly trained shooters to abandon sophisticated firing techniques in a real gunfight and be overcome by instinctive stress responses.

6

Some combat shooting methods, such as the Weaver stance, rely on consciously learned postures and high levels of fine motor skill. These methods require intensive training to master and constant practice if one hopes to use them in the stress of a gunfight.

Based on statistics compiled from actual gunfights, we *know* what the circumstances of a typical gunfight will be. From human nature, we *know* how the body will react to the extreme stress of a life-threatening situation. It makes sense, then, that *practical* combat shooting technique should be based on this real-world knowledge. And so it was . . .

THE DEVELOPMENT OF POINT SHOOTING

During World War II, I was selected as close combat instructor for the secret operatives of the Office of Strategic Services (OSS) and U.S. Army Military Intelligence. Along

In a gunfight, the natural tendency is to focus on the threat, not the weapon's sights. This, along with a lack of time, light, and often opportunity, make sighted fire at close-combat ranges extremely difficult to perform.

with my assignment to this position, I received tasking from the legendary "Wild Bill" Donovan, commander of the OSS, to learn all there was to know about close-quarters combat. Within a period of months, I and the other instructors at Section VIII, Camp Ritchie, Maryland, devised a comprehensive course of instruction which included both unarmed and armed combat methods.

Since the nature of clandestine operations established the handgun as a primary weapon for many covert operatives, we placed special emphasis on the development of an easily learned yet highly effective method of close-quarter combat shooting. We began with a system developed by British commando trainers W.E. Fairbairn and E.A. Sykes while serving with the Shanghai Municipal Police (SMP) during the early twentieth century. While serving with the SMP, Fairbairn and Sykes participated firsthand in scores of gunfights and analyzed the events of literally hundreds of others. Using this real-world knowledge, they devised and implemented a method of combat point shooting which proved to be deadly effective in the stress of a real armed conflict.

I had the pleasure of collaborating with both Fairbairn and Sykes during the development of the instructional program at Section VIII. We adopted their combat-tested technique and taught it to the operatives who passed through our course of instruction. Through a policy of rotating Section VIII instructors to various theaters of operations, we added our own real-world experience and continued to both validate and refine point shooting technique even more. The result was a method of combat shooting which allowed the effective delivery of fire at close range in virtually any light conditions without the use of sights. This method was ultimately taught to thousands of U.S. and allied operatives and successfully applied in combat countless times during the war.

One fact that modern shooters often overlook when considering point shooting as a combat shooting method is that it was developed and refined during periods of actual violent conflict, not on a training range or in a classroom. When Col. Rex Applegate was in charge of training at the Military Intelligence Training Center (MITC) during World War II, it was his policy to rotate MITC instructors to various operational units in different theaters of the war. By participating in actual operational missions, these instructors were able to determine firsthand how well the techniques taught by the MITC worked in the field. When each instructor returned to the MITC at the end of an operational period, his feedback was used to refine and improve the techniques and training methods of the school. If something didn't work, it was abandoned. If it did work, the firsthand combat experiences of the MITC instructors were used to make it work even better.

Unlike many modern gun "gurus," the instructors at MITC were not interested in fame, fortune, or personality cult status. Their only concern was what worked in actual combat. And in their opinion, point shooting worked.

Below is a list describing the credentials and combat experience of just a few of the MITC instructors that served under Colonel Applegate. This list leaves little doubt that these gentlemen knew whereof they spoke.

- An officer who commanded a company in the 1st Ranger Battalion throughout the Africa campaign.
- An officer who had served with an Army Intelligence Section in the African, Sicilian, and Italian campaigns.
- An officer who successfully led 28 patrolling missions against the Germans in North Africa.
- An officer who served with the British Commandos, later with the 1st Ranger Battalion, and then with an infantry division and who was commissioned and decorated on the battlefield for his exploits. He later was sent to the Pacific Theater, where he saw action in New Guinea.

- An officer who served for three months on the Anzio beachhead as a member of regimental and battalion intelligence sections and who was later decorated for his activities in organizing the "battle patrols" of the 3rd Infantry Division.
- A sergeant who had two years of active combat experience in guerrilla warfare against Franco's forces in the Spanish Civil War and was chief of reconnaissance in the Abraham Lincoln Brigade.
- A Greek corporal who had fought with the Albanian, Greek, and Crete campaigns and who later fought in Egypt with the British.
- A corporal who had fought through the Guadalcanal campaign as a member of a battalion intelligence section.
- An officer who had served as a platoon leader and company commander for 2 1/2 years in the southwest Pacific, participating in the Guadalcanal and Bougainville campaigns.
- An officer who had served as a platoon leader during the Aleutians campaigns.
- An officer who served as a company commander in the North African campaign and as an instructor at the Allied Intelligence Training Center in North Africa.
- An officer who served as a liaison officer and in an instructional capacity in the Chinese Army and who participated in active guerrilla warfare behind Japanese lines.

(From *Quick or Dead* by William L. Cassidy, Paladin Press, 1993. Used with permission of the publisher.)

Many of the training methods developed by Applegate and Fairbairn for the OSS were decades ahead of their time. Here Fairbairn accompanies a student during a live-fire exercise in the "House of Horrors," predecessor of modern-day "fun houses" and "kill houses."

In addition to point shooting technique, we also developed a wide variety of training methods to simulate the physical and psychological stress of combat. In retrospect, these training methods were truly revolutionary and predated many so-called modern methods popularized decades later. These included practices such as the use of moving and bobbing targets and a sophisticated indoor range called the "House of Horrors," predecessor of the "fun houses" and "kill houses" currently used by special operations forces and SWAT teams.

Although point shooting was proven countless times in

battle during World War II, the hard-earned lessons of Section VIII and the tremendous advances in combat training methods made there were largely forgotten after the war. Training programs in the military and in law enforcement reverted to bull's-eye-style shooting and other methods that were completely divorced from the realities of combat.

In the decades since World War II, numerous other combat shooting methods have been introduced and popularized. However, unlike point shooting, these techniques were developed primarily for competition-style combat shooting matches rather than actual combat. These matches, no matter how competitive they may be, cannot duplicate the stress of an actual gunfight, where winning is measured in terms of lives, not trophies. More importantly, the effectiveness of point shooting as a combat method for the *average shooter* with limited time and resources for training has been conclusively proven. As the saying goes, if it ain't broke, don't fix it.

THE MECHANICS OF POINT SHOOTING

Point shooting relies on the body's instinctive ability to point at nearby objects with reasonable accuracy. This basic form of eye-hand coordination is an ability that people develop naturally during childhood and refine throughout their lifetimes. It is so natural and well-ingrained that it is not greatly affected *even when the body is suffering reactions to extreme stress.*

The mechanics of instinctive pointing are very simple: the eyes focus on the target, then the arm is raised until the hand breaks the line of sight. By keeping the elbow and wrist locked and raising the arm like a pump handle, a very accurate and consistent alignment of the pointing hand and the line of sight can be achieved.

When a handgun is properly gripped and aligned with the arm, this innate ability to point accurately can be adapted to aim the gun quickly at close-range targets without using the gun's sights. The barrel of the gun simply replaces the point-

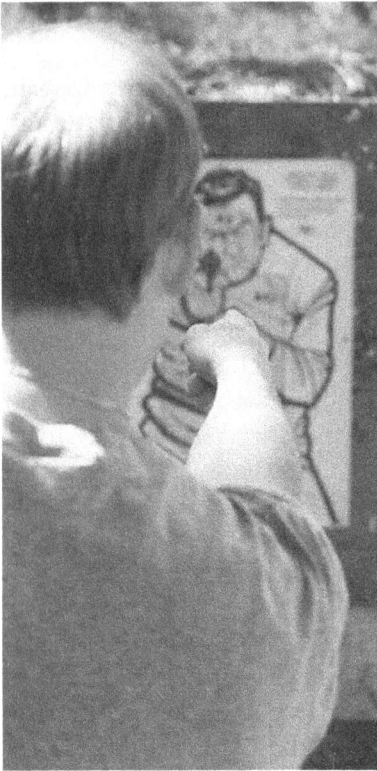

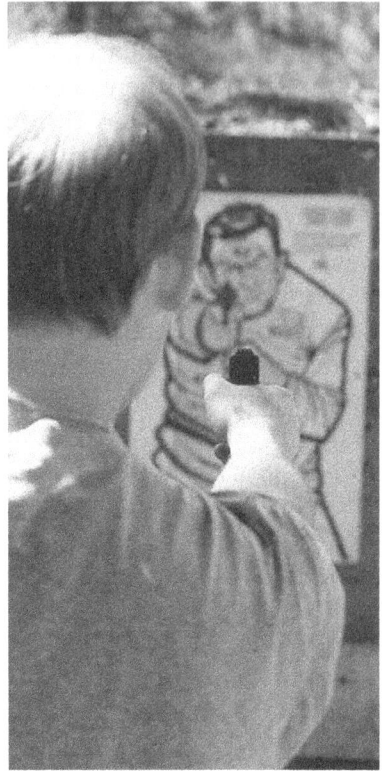

The foundation of point shooting is the body's natural ability to point accurately at close-range targets. By replacing the pointing finger with the barrel of a handgun, this ability can be used to deliver accurate fire without using the weapon's sights.

ing finger. The result is the ability to deliver fire quickly and accurately at close range, in low light, and while experiencing the natural human reactions to extreme stress. As we have already established, this ability is exactly what is needed in a typical gunfight. It is also why point shooting should be the foundation of all combat shooting technique.

COMBAT SHOOTING, DISTANCE, AND TIME

The need to deliver fire in a gunfight depends directly upon how immediate a threat the opponent, or target, is to you. The closer your opponent is, the more immediate the threat and the greater the need to fire your weapon quickly.

In close-quarters situations, the instinctive physical reactions to stress take over, and responses are generally limited to natural, instinctive movements. The ability to perform coordinated movements requiring a high degree of muscle control is seriously impaired.

Conversely, the farther away your opponent is, the less immediate your need to fire. The more time you have before you must fire, the greater your chances of composing yourself and overcoming the instinctive reactions to stress. Overcoming or at least minimizing these instinctive reactions can allow you to utilize a shooting technique that requires a greater degree of coordinated muscle movement. It may also allow you to switch your visual focus from the threat—where it will naturally be in a life-threatening situation—to your weapon's sights.

Realistic combat shooting technique therefore becomes a continuum that ranges from contact distance (so-called "hip shooting") to two-handed, sighted fire. Where *you* fire along this continuum is based on your distance to the target, the perceived threat, and your ability to control the instinctive physical responses to stress.

Statistics have consistently shown that the typical gunfight occurs at close range, in low light, and under conditions of extreme stress. The closer and more imminent the threat, the more immediate your need to fire. In these circumstances, the body's instinctive reactions to stress take over and only natural movements can be performed reliably.

As the range to the threat increases, your need to fire becomes less imminent and your ability to control physical stress increases. In these circumstances, shooting methods which rely on fine motor skills and the use of the weapon's sights may be employed.

COMBAT SHOOTING PHASE 1:
THE BODY POINT

In real gunfights, close distance is the rule, not the exception. Therefore, a practical combat shooting method should begin at the shortest possible range one could encounter. At this range—from contact distance to about 5 feet—the technique of choice is the body point.

THE COMBAT SHOOTING STANCE

The stance for close-quarter combat shooting is based on the instinctive physical reactions to stress. Since these reactions determine the stance you will assume naturally in a high-stress situation, they must logically become the foundation of your combat shooting stance. Again, these instinctive reactions are:

- crouching
- squaring the body with the threat
- visually focusing on the threat
- convulsive muscle contraction

To assume a proper combat shooting stance, first assume a relaxed posture. Take one natural step forward and flex your knees to assume a balanced crouch. You may step with either foot, and it is a good idea to become proficient firing with either foot forward. Your gun hand should be extended downward in front of you at about a 45 degree angle, and your free hand should be held out to the side for balance. In this position, your upper body will be naturally inclined

A proper "combat crouch" with the handgun.

slightly forward in an aggressive posture. Consistent with the instinctive reactions to stress, your eyes will remain open and focused intently on the target.

Practice assuming this stance smoothly at first until both the movement and the resulting posture feel very comfortable and natural. Avoid the temptation to look at your foot placement, and keep your eyes straight ahead on the threat. Once you're comfortable doing the movement slowly, try assuming the stance more suddenly to simulate reacting to an unexpected threat. This practice should be done both from the standing position and while walking. It should also be done on a variety of realistic surfaces such as grass, gravel, and uneven terrain to simulate realistic combat conditions.

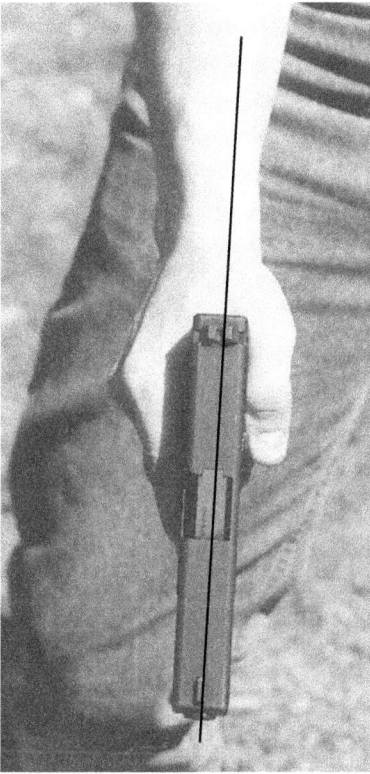

A proper grip is essential to good combat shooting. The handgun should be gripped convulsively with the bore of the gun in line with the forearm.

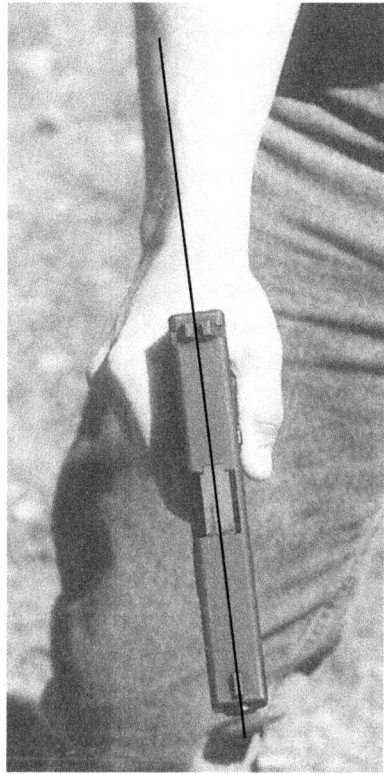

This photo illustrates an improper grip. The bore of the gun is not in line with the forearm.

The combat shooting stance is well balanced and allows you to walk forward and backward or pivot to either flank very naturally. More importantly, it incorporates all the instinctive physical responses to stress and is therefore a sound basis for *all* close-quarter combat shooting.

GRIPPING THE HANDGUN

A proper grip is crucial to effective combat shooting technique. Place the gun firmly in the web of the hand, making sure that the barrel is perfectly aligned with the forearm. This allows the barrel to act as a natural extension of the arm. The web of the hand should be as high as possible on the grip to keep the bore in line with the arm horizontally and reduce muzzle rise during firing.

Grip the gun *convulsively*, contracting the entire hand tightly around the grip stocks. A good convulsive grip is important because it prevents the gun from turning in the hand when the trigger is squeezed, helps control recoil, and ensures weapon retention. Most importantly, however, a convulsive grip is important because *it is how you will naturally grip the gun in the stress of an actual gunfight.*

THE BODY POINT

The first phase of the combat shooting continuum

The body point is used only at extremely close ranges and is the only shooting technique done from below eye level. To perform this technique from a natural stance, step forward into a crouch, place your gun arm elbow against your side, raise your forearm until it is horizontal, and squeeze the trigger with a convulsive contraction of the entire hand.

is the body point. This technique is designed to be used only at extremely close range. To perform this technique, first assume an instinctive crouch. Then, press your gun-arm elbow tightly against your ribs and raise your forearm straight ahead, leveling it at the target. If the gun is gripped properly, it will naturally point where the forearm does. Once the gun is on line with the target, contract your entire hand in a uniform squeeze. The pressure of the index finger on the trigger will fire the gun, while the pressure of the other fingers will keep it firm in the hand and steady on target.

At ranges from contact distance out to about five feet, this firing technique is extremely effective. Aiming in this manner

The body point is used only at near-contact ranges. Note that the non-firing hand may be used to strike or deflect blows before the shot is delivered.

To engage multiple targets with the body point, fire at the first target, then pivot smoothly on the balls of the feet to face the next target.

yields acceptable close-range accuracy, and the bent arm posture protects the gun hand from being grabbed. This technique can also be combined with a strike or push with the free hand to keep your opponent from closing, and it may be executed directly from a strong-side draw. Note that if the body point is employed with any type of free-hand warding or striking action, you must be extremely careful to avoid letting your hand cross in front of the muzzle of your gun.

To shift your fire laterally when using the body point, assume a proper stance, then pivot on the balls of your feet, turning your body like a turret until you face the target directly. Keep your gun-arm elbow pressed tightly to your side and turn your entire body. This allows for more accurate fire than simply moving your arm, and it keeps your body squared with the threat.

COMBAT SHOOTING CONTINUUM PHASE 2: POINT SHOOTING

A s we have already seen, the vast majority of gunfights occur at ranges of about 5–20 feet. Shooting techniques that have proven effective at this range should therefore be the foundation of *all* combat shooting training and should be regarded as *the single most important life-saving skill for individuals armed with handguns.*

On the practice range, highly skilled shooters can successfully use two-handed aimed fire techniques to hit quickly and consistently at typical gunfight distances. However, in the stress of a gunfight against targets that are shooting back, even highly skilled shooters are often overcome by instinctive reactions and revert to unsighted point firing. *Since point shooting is the natural tendency in a real gunfight, it should be the rule in realistic combat shooting training.*

Additionally, we must remember that most gunfights occur during the hours of darkness in lighting conditions that make sighted shots difficult, if not impossible. Even if you do succeed in making a sighted shot in low light, the odds of making a second are slim once your night vision has been destroyed or at least impaired by the light of your own or your opponent's muzzle flash.

POINT SHOOTING

As with the body point method, a proper grip on the

handgun is critical to effective point shooting. Again, you must place the gun firmly into the web of the hand, with the hand as high as possible on the grip. The barrel should be aligned with the forearm so it becomes a natural extension of the arm. This is critical and must be practiced until it can be assumed automatically the instant you grasp or draw the handgun. Once you are able to grip the gun correctly, practice doing so with a strong, convulsive squeeze of the hand. In a gunfight, you will naturally grip the gun in this manner, so you might as well get used to doing it now.

It should be noted that, when firing semiautomatic pistols, a strong, convulsive grip also eliminates the problem of "wimp wristing." This is a stoppage that occurs when a weak grip allows the pistol to move rearward excessively in recoil. The frame of the pistol moves simultaneously with the recoiling slide, preventing the slide from cycling fully and chambering a fresh round.

Point shooting is the most important life-saving skill for individuals armed with handguns. To perform this technique from a neutral posture, first take one natural step forward and bend your knees to assume an aggressive crouch. Your gun arm, wrist, and elbow should be locked and your arm should be extended at about a 45-degree angle to the ground. Your free hand is held to the side for balance. With your eyes focused on the target, raise your gun arm until the gun reaches your line of sight, then contract your hand uniformly to fire.

The starting position for point shooting is once again the instinctive crouch. As described earlier, your feet should be spread a comfortable distance apart, with one foot a natural step forward of the other. Bend your knees and incline your upper body slightly forward to establish a balanced yet aggressive posture. Your gun arm should be extended straight downward at about a 45 degree angle, with the wrist and elbow locked, while your free arm should be extended loosely to the side for balance. In this position, your shoulders will remain squared with the target. Consistent with the instinctive reactions to stress, your eyes will remain open and focused straight ahead on the threat.

Standing in the crouch with your gun gripped properly, raise your arm until the gun reaches your line of sight to the

target. During this move-
ment, the wrist and elbow
remain *locked* and the arm is
raised from the shoulder with
an action resembling a pump
handle. The gun should stop
immediately once it reaches
the line of sight, but your
focus should remain on the
target, where it will naturally
be in the stress of a gunfight.
Do not attempt to shift your
focus to the gun's sights or
establish a sight picture.

Lower the gun arm to the
45 degree angle starting position and repeat the move-
ment again slowly. Continue
to practice this movement
slowly, paying special attention to the feeling of your
locked wrist and elbow and
the smooth rise of the gun to
eye level. Again, *do not* look at
your arm but maintain your
focus on the target. Control
the movement of your arm so
the gun stops immediately
when it reaches eye level and

Point shooting technique as seen from the side. Note the angle of the gun arm in the ready position and the fact that the gun is brought to eye level before firing.

does not drift up past the target. As you become more comfortable with the movement, increase your speed gradually.

Now that you are comfortable with raising your gun to
eye level, it's time to combine the movement with dry fire.
*Before beginning this practice, clear and double check your gun to
ensure it is unloaded, and always practice facing a safe direction.*

Assume a good instinctive crouch and raise your gun with the wrist and elbow locked as before. This time, as soon as the gun reaches eye level, contract your entire hand convulsively to squeeze the trigger. Do not pause before you squeeze the trigger or attempt to establish a sight picture. Squeeze the trigger the instant the gun reaches your line of sight.

With a little practice, you'll find that if the gun is gripped properly and the arm is raised to the line of vision, the gun will naturally point where you are looking *without having to use the sights*. As such, although point shooting is not sighted fire, it most certainly is *aimed* fire.

POINT SHOOTING LIVE FIRE

Once you are comfortable with dry fire practice, it's time

These close-up photos show the movement of the gun arm as it is raised. The wrist and elbow remain locked throughout the movement, which may be compared to the raising of a pump handle.

This photo clearly shows that the gun is raised fully to the line of sight before firing.

to fire some rounds. Begin at relatively close range (about 10 feet) with single rounds. As with the dry fire drills, as soon as the gun reaches eye level, contract your entire hand tightly to fire. Don't worry about fine trigger control or precise alignment of the gun. The uniform contraction of the hand will keep the gun positioned properly. You will also find that it is extremely effective in controlling recoil since all the fingers of the hand contract together at the instant the round is fired.

If you are doing everything right, you will find that you have no difficulty consistently putting your rounds into the center mass of the target. If your hits are not going where your eyes are focused, check your grip on the gun to make sure the barrel is aligned with your forearm, and make a conscious effort to keep your wrist and elbow locked.

Keeping the wrist and elbow locked and *raising* the gun to

This sequence utilizing Hornady Vector ammunition shows point shooting in action. Photo one shows the gun as it reaches eye level, photo two shows the round impacting the center plate of the target, and photo three shows the center plate reacting to the bullet strike.

One of the most common errors made in handgun shooting is to shove the gun forcefully at the target rather than raising it. The result of this action is that the momentum of the gun causes the muzzle to dip when the hand stops at full extension.

the line of sight is the key to successful point shooting. One common mistake, especially in stressful situations, is for the shooter to shove the pistol forward rather than raise it. Shoving the pistol causes the wrist to flex at full extension and the muzzle of the gun to dip downward. The often seen raised-gun ready position, where the gun is held muzzle up near the shoulder before it is aimed, makes this phenomenon very likely under stressful conditions. In fact, *most of the missed shots that occur in actual gunfights can be attributed to the shooter shoving the gun forward from this position.*

Shoving a gun forward to fire it will invariably result in low hits because the muzzle will dip when the arm reaches full extension. The degree to which a gun's muzzle dips when the weapon is shoved varies, however, from one handgun design to another. The design factor that determines muzzle dip is the angle between the grip and the barrel. The closer this angle is to 90 degrees, the more the muzzle will dip when the gun is shoved. Conversely, designs with grip angles well in excess of 90 degrees are much less susceptible to muzzle dip. Handguns of this type typically feel very comfortable in the hand and point very naturally because the grip lies across the palm of the hand rather than perpendicular to it. Additionally, handguns with large grip-to-bore angles allow the hand to be placed closer to the bore line, making recoil control easier and greatly reducing muzzle rise during firing.

To demonstrate the effect grip angle has when shoving a handgun, we fitted a Model 1911 .45 automatic with a laser sight. This gun has a grip-to-bore angle of 107 degrees. Standing approximately 15 feet from a silhouette target with the laser turned on, we had several shooters take aim at the target's head by shoving the gun from shoulder level. Invariably the laser dot, and consequently the point of impact of any rounds fired, dropped well below the intended target when the pistol was shoved. In most cases, intended head shots became shots to the groin or legs.

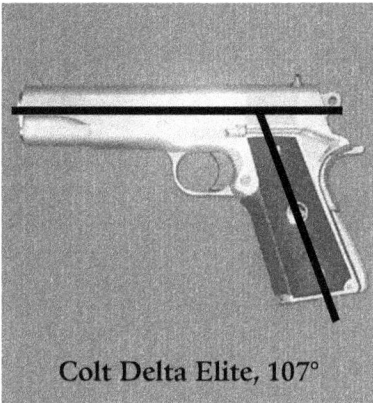

Colt Delta Elite, 107°

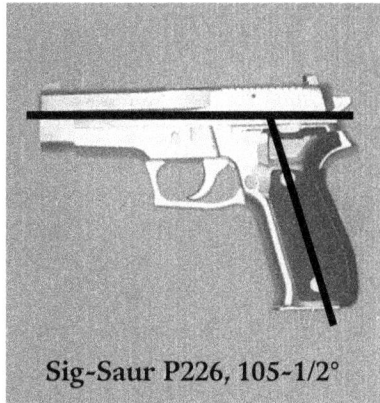

Sig-Saur P226, 105-1/2°

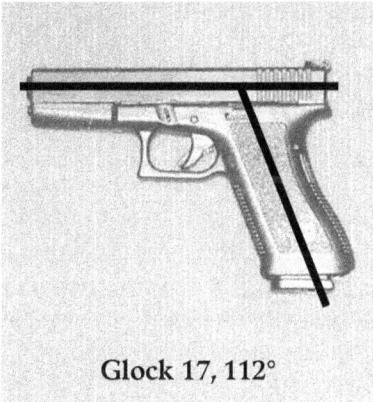

Glock 17, 112°

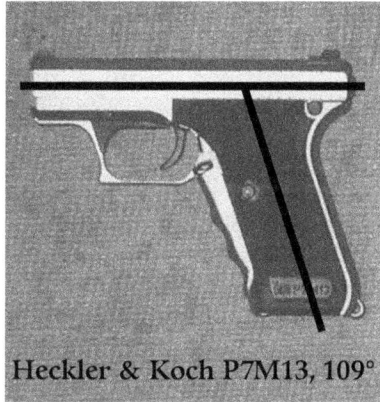

Heckler & Koch P7M13, 109°

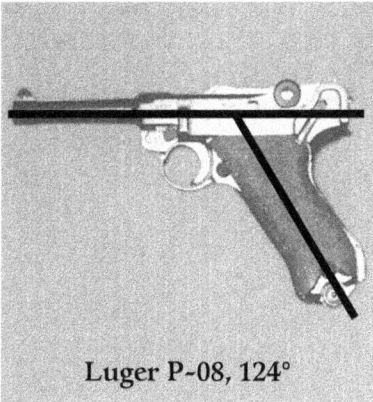

Luger P-08, 124°

The angle of the grip to the bore of a handgun determines how radically its muzzle will dip if shoved at the target. The closer this angle is to 90 degrees, the more the muzzle will dip. Guns with greater grip angles dip much less and tend to point much more naturally.

When the shooters saw the laser dot drop and realized the effect shoving the pistol had on their aim, most tried to compensate by raising the pistol back up to bring it on target. Needless to say, these corrections were awkward and imprecise and would have made accurate shooting very difficult. More importantly, corrections of this type would waste precious time in a gunfight—time that could easily mean the difference between living or dying.

For comparison, we fitted the frame of a Whitney Wolverine pistol with a laser sight. This pistol features a grip-to-barrel angle of 123 degrees and has a distinctly different feel in the hand. In identical test circumstances, shooters found that the laser dot of this pistol dropped only slightly, even when shoved forcefully at the target. Most intended head shots remained head shots, and even very violent shoves caused the laser dot to dip no lower than throat or shoulder level.

The effects of shoving a pistol were even more apparent during actual firing. Test firings of a number of popular service pistols at a range of 15 feet revealed that shoving them forcefully from the shoulder caused intended center mass hits to impact *3 to 4 feet low.* In some cases, shoving these pistols caused rounds to impact *in the dirt in front of the target.*

In contrast, rounds fired from a German P-08 Luger by the same shooters using the same ammunition dropped an average of only 1 foot at the same range. The P-08 has a grip-to-barrel angle of 134 degrees. The average grip angle of the modern service pistols tested was 108 degrees degrees.

Clearly, handgun designs with greater grip-to-barrel angles point more naturally and are less susceptible to muzzle dip than designs with grip angles close to 90 degrees. Choosing a handgun design that points well will therefore help increase your chances of hitting the target. However, the key to successful point shooting with *any* handgun is still raising your gun arm with wrist and elbow locked.

Here we see point shooting in use on a steel reactive target. Note the shell casing in the air and the center plate of the target in motion from a well-placed hit.

Remember the golden rule of point shooting: *thou shalt not shove thy handgun.*

With a little practice at proper point shooting technique, you should find that you have no difficulty keeping single rounds on target. Once you've developed this ability, you should establish the habit of firing two quick rounds in succession each time you raise your gun to fire. *The idea of this tactic is not to try to group the shots closely as in competition shooting, but simply to improve your chances of stopping the attack by hitting the target twice.* Keep in mind that it is natural for the recoil of the handgun to disperse your shots somewhat, and resist the temptation to pause between rounds to refine your aim.

With proper technique and a good convulsive grip, most

average shooters quickly develop the ability to keep their hits 6 to 8 inches apart at ranges out to about 20–30 feet. This standard of combat accuracy will enable you to consistently hit head-sized targets at typical combat distances. It will also enable you to effectively target your opponent's vital organs in a gunfight.

POINT SHOOTING WITH POCKET PISTOLS

One great advantage that point shooting offers over shooting methods that rely exclusively on sighted fire is that it works well with guns that have small sights or no sights at all. Despite the fact that many so-called gun experts scoff at the use of small caliber "pocket pistols" for defensive purposes, these pistols are actually the most commonly carried defensive handguns. Most legally armed citizens and many off-duty law enforcement officers prefer these guns to their big-bore brothers because they are convenient to carry and easily concealed. These small guns are also popular backup weapons for on-duty officers carrying large caliber sidearms.

Small pistols typically have diminutive sights that are extremely difficult to see even in good light. Some designs, such as the popular Seecamp and AMT Backup models, are either totally devoid of sights or have only a shallow sighting groove milled into the top of the slide. Sighted fire techniques are therefore extremely difficult to employ with these pistols, especially under the stress of combat. This is particularly true of sighted fire methods that require a two-handed hold on the pistol, since most of these guns are barely substantial enough to grip with one hand.

Since point shooting does not rely on the use of sights, it works as well with pocket pistols as it does with larger handguns. Moreover, the fact that it requires only one hand makes it instantly adaptable to *any* handgun regardless of grip size or configuration.

POINT SHOOTING IN LOW LIGHT

The greatest advantage of point shooting, however, is that it allows the accurate delivery of fire in low light. As we have already established, most gunfights occur during the hours of darkness or semidarkness. In these conditions, even if you had time to look at your sights, you probably couldn't see them well enough to utilize them.

To illustrate the effectiveness of point shooting in low-light conditions, we covered the sights of a pistol with tape and took a trip to the range at dusk. In subdued light, shooters had no difficulty hitting the target using point shooting technique. As darkness continued to fall, our shooters were still able to quickly and consistently score hits on target. Not until it became so dark that the targets were no longer easily discernible with the human eye were our shooters forced to cease fire.

In low light, it is natural for one to have to focus intently on a target in order to see it clearly. This intense focus is no different than that experienced in the stress of a gunfight and therefore remains perfectly compatible with point shooting methodology. Again, no adjustment or modification of the basic technique is necessary to make it effective in different circumstances. The bottom line remains that at typical gunfight distances, if you can see your target, you can hit it with point shooting.

In gunfights that occur in extreme darkness, often the only aiming point available to the shooter is the opponent's muzzle flash. Under these conditions, point shooting enjoys a decided advantage over sighted fire. When shooting at the fleeting image of a muzzle flash, the lack of a distinct and constant aiming point makes using traditional sighted fire practically impossible. Typically, the shooter will instinctively shove or push the weapon toward the muzzle flash, causing the muzzle to dip and the shot to impact low. Sighted fire shooters also often "snap" or "jerk" their shots when firing at muz-

One great tactical advantage of point shooting is that it allows the non-gun hand to be used for such common tasks as holding a flashlight or opening a door.

zle flashes in a desperate effort to use the momentary illumination to align their sights.

Point shooters, on the contrary, need only maintain their focus on the spot where the muzzle flash appeared, raise the gun to eye level with proper point shooting technique, and fire. The effectiveness of point shooting in conditions of extreme darkness should, alone, be reason enough to incorporate point shooting training into all mil-

Point shooting also enables the shooter to fire effectively from disadvantaged positions.

itary and law enforcement training programs.

In many low-light situations, especially in law enforcement, hand-held flashlights are employed in conjunction with handguns. For shooters trained in two-handed sighted fire techniques, this poses a problem since they must figure out a way to hold both the flashlight and the handgun while maintaining their two-handed firing posture. In most cases, the result is an unnatural contortion that comes apart as soon as the first round is fired. These contorted postures also place the flashlight directly in front of the shooter's body, making him easily targeted by bad guys.

Once again, the fact that point shooting is a one-handed shooting method naturally solves these tactical problems. First of all, the non-gun hand is free to hold and operate the flashlight naturally without resorting to complicated postures. Better yet, this hand is free to hold the flashlight in any position, including postures with the arm fully extended away from the shooter's body. This makes it possible to shine the light at different angles while keeping the gun hand in a natural ready position. It also allows the shooter to use the light while taking full advantage of cover and concealment. Best of all, the fact that the light is not held in front of his body greatly reduces the chances that he will be hit by hostile fire aimed at the source of the light.

WEAK-HAND SHOOTING

In combat, it is quite possible that your strong arm or hand may be injured, wounded, or immobilized, forcing you to shift your handgun to your weak hand. In such situations, point shooting once again shines.

First of all, since point shooting is fundamentally a one-handed method, it is easily and naturally performed equally well with either hand. Changing suddenly from a two-handed shooting method with the gun held in the strong hand to a one-handed weak-hand hold is very dramatic. The dissimilar nature of the two methods makes it necessary to learn and master them separately. Conversely, point shooting is essentially the same regardless of which hand holds the gun.

Additionally, since both eyes remain open and focused on the target in point shooting, the weak-hand firing process is not complicated by forcing the shooter to "switch gears" visually. With sighted fire methods, although both eyes typically remain open, one eye, called the master eye, is primarily responsible for focusing on the sights. In most shooters, the master eye corresponds to their strong hand, i.e. right-handed shooters typically have right master eyes. In normal strong-hand shooting, this makes it natural for them to sight down their strong arm to the gun's sights. In weak-hand shooting, however, they must hold their gun hand across their body in front of their master eye and consciously adjust their vision to achieve a sight picture. This is difficult, unnatural, and wastes precious time when rounds are flying in your direction.

MOBILITY AND POINT SHOOTING

Multiple targets can be engaged easily with point shoot-

To engage laterally spaced targets with point shooting, fire on the first target, then lower your gun arm to the ready position as you pivot smoothly on the balls of your feet to face the second target. As you complete the pivot (page 44), raise your gun arm to eye level and fire again.

ing by using the same pivoting movement utilized with the body point technique. From the instinctive crouch, pivot smoothly on the balls of your feet until your shoulders are squared with your target, then raise your hand and fire. Lower your hand while pivoting to the next target and raise it again to fire. This will ensure that your body is aligned with the target before you attempt to squeeze the trigger.

One common mistake when engaging multiple targets or targets that appear on your flank is to leave your feet in place and swing your gun arm. By swinging your arm without moving your body, you lose the ability to accurately index on your target by squaring your shoulders to it. It is very difficult to control independent arm movements precisely and nearly impossible to use them to aim accurately or consistently.

Another common mistake is to reposition your feet by jumping rather than pivoting. Once again, such a movement is very difficult to control, and it is unlikely that you'll ever be able to land with your feet in the same place twice.

Curiously, a version of this technique was actually adopted and taught for many years by the FBI. Known as the FBI Crouch, this technique called for the gun hand to be held low near the hip and for the

shooter to engage targets by jumping to change direction. Needless to say, it is not very effective.

This photo shows the disadvantage of two-handed shooting methods when engaging targets to your flanks. A two-handed grip locks the upper torso and makes smooth pivoting impossible.

Smooth pivoting on the balls of your feet allows you to quickly face multiple targets while maintaining complete control of your movement. It is also the best way to adapt to irregularities in the terrain, obstacles, and other detritus without losing your balance. To prove this to yourself and to develop the ability to maneuver smoothly in any situation, you should practice this movement regularly on different surfaces and types of terrain. You should also practice while wearing different styles of footwear to ensure that you will be able to move effectively no matter how you're dressed.

Pivoting can also be utilized to engage targets to your flanks while moving. Begin by assuming an instinctive crouch, then walking while maintaining this same basic posture. Take natural steps, and keep your weight evenly balanced over both feet while maintaining an aggressive upper body posture. Done correctly, you should bear a striking resemblance to Groucho Marx. When a target presents itself, stop your forward motion and pivot smoothly to face it. In most cases, you will be able to pivot immediately, but some pivots may require a short extra step to face the target squarely.

For example, if you are stepping with your left foot as a target appears to your right flank, you can pivot immediately as your left foot touches the ground. If you are stepping with

Point shooting is ideal for engaging targets to your flanks while you are on the move. Begin by walking while remaining crouched and keeping your gun arm in the ready position. When a target presents itself to your flank, plant your foot and smoothly pivot on the balls of your feet to face it. Then, raise your gun arm and fire.

your right foot, however, it may be necessary to take a small additional step with your left foot to complete the pivot.

As with the stationary pivot, practice walking pivots on different types of terrain until you can pivot reflexively in either direction and fire with either foot forward. To appreciate the ease with which you can pivot in a point shooting stance, try doing walking pivots while holding your gun in a two-handed grip. Two-handed shooting methods typically lock the upper body in a predetermined firing position, making pivoting to your flanks difficult and unnatural.

THE FAIRBAIRN/SYKES TECHNIQUE

One variation of the point shooting technique described herein is the original method developed and practiced by

Fairbairn and Sykes. In this method, the gun is raised to chest level and held directly in front of the centerline of the body. Since this technique does not take full advantage of the body's instinctive eye-hand coordination, it is inherently less accurate than bringing the hand to eye level. The point shooting method you have just learned is an improvement upon the original Fairbairn/Sykes method that was found to be even more effective in actual shooting incidents. It is also easier to learn and master and is therefore preferred to the original technique.

47

COMBAT SHOOTING
CONTINUUM PHASE 3:
TWO-HANDED
POINT SHOOTING

As we have already established, distance plays a critical part in combat shooting. At close range, extreme stress and the overwhelming need for a swift and deadly response make the natural movements of the body point and one-handed point shooting the only viable techniques for the average shooter.

At greater distances, where the threat of a hostile opponent is less imminent and you have slightly more time before you must fire, it is possible to use your free hand to reinforce and help steady your gun hand. Initially, this should be practiced by first executing a standard one-handed point shooting response, then raising the free hand to form a two-handed grip. This further reinforces the standard point shooting technique and allows you to fire before establishing a two-handed grip if necessary.

To assume a correct two-handed grip, wrap the fingers of your free hand tightly around the front of your gun hand. In addition to steadying your hold on the gun and helping to control recoil, this technique is useful for individuals with small hands who have difficulty gripping large pistols solidly with only one hand. This is often the case with police departments that adopt a standard sidearm for all officers without concern for individual variances in stature and hand size.

At greater ranges, a two-handed grip may be employed to steady the gun. This grip is also useful for shooters with small hands who must fire handguns with large grips. To assume a two-handed grip after taking aim by point shooting, simply raise the free hand and wrap it around the front of the gun hand to apply a firm pulling pressure. The locked wrist and elbow of the gun arm resists this pull to create a dynamic tension between the hands. Both elbows remain straight to create a steady triangle support for the gun.

The Isosceles Stance

If you are at or near the limits of your one-handed point shooting abilities and have the presence of mind to do so, you may establish a two-handed grip *before* raising your pistol. Keeping your elbows straight and your eyes focused on the threat, raise your arms together until the gun reaches your line of sight. The other movements of the body, including assuming a crouch and squaring the shoulders with the threat, remain consistent with the instinctive reactions to stress. This method of two-handed shooting is commonly known as the isosceles stance.

You will probably see some individuals assume an isosceles stance by spreading their feet about two shoulder widths apart and squatting deeply, with their shoulders squared to the target. Avoid this practice because this foot positioning does not provide the mobility and balance of the instinctive crouch.

In its basic form, the isosceles stance is nothing more than point shooting with a slightly more stable platform. Since both arms are held straight, the gun remains aligned with the vertical centerline of the body. When raised to the line of sight, it will naturally point where the eyes are looking. Your focus stays where it belongs, on the target, and all other aspects of your stance and posture remain the same. The only difference is that you are now pointing with two hands instead of one.

If you have sufficient time and your non-gun hand is not being used to open a door, operate a flashlight, or perform any other tactical movement, a two-handed hold can be used to reinforce your point shooting technique and extend your effective range. However, with the exception of individuals with small hands, it should generally be used as an adjunct to, not a replacement for, one-handed point shooting.

To assume an isosceles stance directly from a crouched ready position, wrap the fingers of the free hand around the front of the gun hand and lock both elbows. The free hand exerts a firm pull while the gun hand pushes. Raise both arms together until the gun reaches eye level, then fire. If time and light conditions allow, the gun's sights may be employed.

COMBAT SHOOTING CONTINUUM PHASE 4: TWO-HANDED SIGHTED FIRE

I f the threat you face is not imminent, you are able to control your reactions to stress sufficiently, can afford to take your focus off the threat for a split second, *and* have sufficient light to see your sights, two-handed *sighted* fire may be used. The most basic technique for this begins with the two-handed Isosceles point shooting stance. Once you have raised the gun to eye level, shift your focus from the target to your sights. The front sight is the most important point of focus and should be visually superimposed on the target. This is often referred to as a flash front sight picture.

If you have achieved sufficient mastery of one- and two-handed point shooting methods, you will find that very little or no adjustment of your aim is necessary when you shift your focus to the weapon's sights. This is as it should be, since good point shooting technique will naturally aim your weapon at the target. At the risk of being redundant, point shooting is not *sighted* fire, but it most certainly is *aimed* fire.

THE WEAVER STANCE

A more sophisticated method of two-handed aimed fire is the Weaver stance. With this method, the shoulders and feet are turned at about a 45 degree angle to the target, and the body is more upright. The elbow of the supporting arm is bent and the supporting hand pulled back while the gun

The Weaver stance requires a high degree of fine motor skill. To assume this stance, begin in an isosceles stance with the feet placed in a weak-side-forward instinctive crouch. Straighten your knees to stand more erect and flex the elbow of your free arm while keeping your gun arm straight or nearly so. This will turn your torso so it faces about 45 degrees to the target and will create a strong push-pull tension between your hands.

hand, which is to the rear, is pushed forward. This creates an isometric tension which helps control recoil and allows for rapid follow-up shots.

Although the Weaver stance provides an extremely stable firing platform and helps tame the recoil of powerful handguns for rapid follow-up shots, the finely coordinated body positioning of this stance is obviously contrary to the instinctive physical responses to stress. This technique therefore requires considerable practice to master and continuous reinforcement training if you hope to be able to use it in a gunfight. Like all sighted fire techniques, it also depends upon the availability of sufficient levels of light to be effective. For the average shooter, it is therefore best reserved for use during long-range combative encounters when the shooter has the time and distance to consciously adjust his tactics.

Once again, *realistic* combat shooting technique is a continuum that ranges from extreme close range body point firing to two-handed sighted fire techniques. Where *you* fire along this continuum depends upon your level of training, the circumstances of the violent encounter, and, most importantly, your ability to function while experiencing the instinctive physical reactions to life-threatening stress. You may proceed through several stages before you fire, or you may decide upon your firing technique immediately. Remember, however, that most handgun encounters occur suddenly, in low light, and at close range. In these conditions, there is no time for indecision or error.

For the average shooter facing the conditions of the typical gunfight, point shooting remains the most natural and effective combat shooting technique. Although practice in aimed fire methods is useful, it should not be done at the expense of point shooting training. *Shooters trained exclusively in two-handed sighted fire techniques are, in reality, only half trained for the conditions of the battlefield or streets.*

Realistic combat shooting technique is a continuum which ranges from extreme close range body point firing to two-handed sighted fire techniques. Where you fire along this continuum depends upon your level of training, the circumstances of the violent encounter, and, most importantly, your ability to function while experiencing the instinctive physical reactions to life-threatening stress.

SHOOTING FROM
THE DRAW

In some cases, you will not be able to have your gun in hand before a fight begins, and you will be forced to deliver fire immediately after drawing from a duty holster or concealed carry position. If you carry a handgun on your person, you must therefore practice drawing and firing it quickly.

All of the shooting techniques described previously in this book may be effectively employed directly from the draw stroke. The secret of doing this is to integrate the draw stroke with the instinctive combat crouch and proceed directly from "clearing leather" to the appropriate preparatory position for the shooting technique you choose to employ.

The easiest way to accomplish this is to position your holster on the gun-hand side of the body. A holster constructed to present the handgun with a slight forward cant is preferred, since it allows the gun to clear the holster smoothly when the body assumes an aggressive forward crouch.

Learning to draw a handgun should be done slowly and carefully. Practice first with an empty gun and pay particular attention to the positioning of the trigger finger during the draw. *Never allow the trigger finger to touch the trigger until the gun has cleared the holster and is pointed away from your body.* Do not attempt a draw with a loaded handgun until you have perfected your technique with an unloaded one.

To draw the handgun, assume a combat crouch and reach back to establish a secure grip on the gun while keeping your trigger finger straight along the frame. Your grip on the gun should be the same as your firing grip, bisecting the angle of

This sequence shows point shooting employed from a draw. Beginning from a natural stance, the shooter first assumes an aggressive crouch and obtains a solid grip on the gun, making sure to keep the trigger finger straight and out of the trigger guard. To draw the gun, raise your elbow while allowing your wrist to bend. This clears the gun from the holster with less arm movement than a straight-wristed draw. Once the gun has cleared leather, extend the gun arm slightly downward and lock the wrist and elbow. Finally, raise the gun to eye level and fire.

the web of your thumb to position the bore on line with the forearm. This way, you avoid having to shift or adjust your grip before firing. Provided that you have a well-designed combat holster, you should be able to release any retaining straps or thumb breaks as you establish your grip. Now, raise your elbow and simultaneously allow your wrist to bend inward slightly to remove the gun from the holster. The flexing of the wrist is important during the draw since it allows you to clear the gun from the holster with much less arm movement than a straight-wristed draw.

Once the gun has cleared the holster, and while keeping the trigger finger straight, lower your elbow and straighten and lock your wrist so the muzzle points downward just ahead of your hip. From this position, you may continue to perform any of the firing techniques described earlier. To shoot with the body point, raise your forearm until it is parallel to the ground and tuck your elbow in to your side. For point shooting, continue to extend your arm downward until the elbow is locked at the proper 45 degree ready position. Then raise your arm and fire.

If the threat is not imminent and you choose to assume a two-handed grip before raising your gun, be extremely careful of the position of your support hand. If your support hand moves too quickly, it could pass in front of the muzzle as the gun is drawn, an unsafe and potentially very unpleasant act. To avoid this, practice extending your gun arm completely in the accepted point shooting fashion before adding the support hand. This way, the muzzle will always extend safely beyond the support hand.

As with the individual firing techniques, care must be taken not to shove the handgun when firing from a draw. You may be tempted to push the gun straight forward after it clears the holster to get it on target faster. The result, however, is the same as shoving the pistol: the muzzle will dip and your shots will hit low. Only by extending the gun arm, then *raising* it to eye level, will your shots be consistently on target. The difference in speed of presentation is negligible, and as the saying goes, you can't miss fast enough to win a gunfight.

Drawing the pistol from other carry positions, such as shoulder holsters, crossdraw holsters, etc., is typically not as fast or as natural as a strong-side draw. However, the shooting techniques described in this book can still be effectively employed when drawing from these carries. Again, the goal is to establish a proper shooting grip before executing the draw, then move to the preparatory position for the shooting technique you choose to use with the least amount of wasted motion possible while keeping the trigger finger straight and off the trigger and without allowing the muzzle to point at any portion of your body. As with a strong-side draw, the secret to developing a good draw from other carry positions is to go slowly and gradually build speed as your muscle memory improves.

POINT SHOOTING vs.
THE STATE OF THE ART

Point shooting has been criticized by some so-called authorities as outdated and only capable of providing inaccurate, unaimed fire. However, statistics compiled from shooting incidents on today's streets continue to show that it is what most people instinctively rely on in the stress of a gunfight, regardless of their formal training in sighted fire techniques.

One of the latest developments in combat shooting technology is the use of gun-mounted flashlights and laser sights that illuminate or project a visible dot on the target marking the intended point of bullet impact. The development and use of these sights is significant because it acknowledges that in the heat of a gunfight, you will in fact look at the target, *not* your weapon's sights. It also acknowledges that most gunfights occur in low-light conditions in which conventional sights are useless.

In order to successfully use a laser-sighted weapon with two-handed sighted fire techniques, these techniques must be modified to allow the shooter to maintain his focus on the target. For example, when a pistol is held in a proper Weaver grip with a flash front sight picture, the laser dot actually disappears behind the weapon's sights. However, when used with one- or two-handed point shooting technique, the laser dot not only remains visible to the shooter, it is automatically aligned on the target.

Another technological advance that is affecting special operations forces and SWAT teams is the widespread availability of night vision devices (NVDs). These light-amplification devices allow soldiers and law-enforcement officers to conduct offensive actions in very low-light conditions while maintaining near normal vision. Some styles of NVDs can be strapped to the head, allowing the operator complete freedom of movement and the ability to employ various weapons. This obviously provides a great tactical advantage in close combat situations; however, it also affects the way in which the operator's weapon can be employed. The optics of the NVD alter the operator's view and, once again, make traditional sighted fire methods difficult or impossible.

In the same way, the increasing popularity of protective masks and the use of tear gas, "flash-bang" explosive distraction devices, and obscuring smoke in tactical scenarios severely limits the effectiveness of two-handed sighted fire techniques. The answer to these problems? You guessed it: point shooting.

Die-hard advocates of sighted fire methods will continue to cling to their techniques, modifying them when necessary to make them effective in different tactical arenas. In most cases, these modifications ultimately result in recreating point shooting technique: the ability to instinctively point your weapon while maintaining your focus on the threat. Point shooting does this with or without lasers, night sights and other electronic gadgets and has been winning gunfights for more than 75 years. So why not learn one technique that you can bet your life on no matter what weapon or what attachments you're using?

TRAINING TIPS

L ike any other skill, you must practice point shooting if you wish to become proficient at it. The fact that point shooting is sometimes mistakenly referred to as "instinctive" shooting implies that it is an inherent ability that requires no practice. This is simply not the case.

Compared to other shooting methods, however, point shooting is very readily learned and requires very little training time to develop functional proficiency. Better yet, once learned, point shooting is very easily retained, even with little or no follow-on training. Nevertheless, if you include point shooting technique in your arsenal of defensive skills, it is highly recommended that you practice it regularly. It is also recommended that this practice be done realistically and in a way that challenges you to perform under stress. This not only refines your physical abilities, it gives you the confidence and psychological toughness to apply them in combat.

There are many different training exercises you can do to develop your point shooting skills. The simplest and most basic of these is dry firing. Dry firing is nothing more than practicing your firing technique with an unloaded gun. While this may seem pointless, it is actually an excellent method of polishing your technique without the need for a range facility or the expense of large amounts of ammunition.

To engage in dry fire practice, first clear and check your weapon to ensure it is unloaded, *then check it again.*

Remember that every gun is to be considered loaded until you *personally* clear and check it. Once you are sure you have an empty gun, select a target and slowly practice your technique. Concentrate intently as if you were firing live rounds, and do not allow yourself to become relaxed or lackadaisical in your practice. As you become more proficient, increase your speed and try firing in response to an outside stimulus such as a training partner blowing a whistle or giving a verbal command.

In addition to stationary dry firing, you can also practice maneuvering and engaging targets on the move. This helps you develop your ability to walk and pivot smoothly while point shooting. A great way to practice this is to move through your house or apartment selecting targets at random while you pivot and fire. Make full use of cover and concealment as you do this.

To make it really interesting, mount some silhouette targets to cardboard backings and have a friend place them around your house without telling you their locations. Then do a dry-fire house clearing to locate and dispatch all the targets. To add variety, you can even mix in shoot and no-shoot targets.

Despite its training value, dry firing allows for only limited skill development. For obvious reasons, it is more realistic when done with revolvers or double-action-only automatics than with single action guns. It also doesn't tell you exactly where your shots would have impacted.

The next step up from dry fire is to practice with a laser sighting attachment. This allows you to see where your rounds would impact, yet saves you money on ammo and range fees since you can practice with it anywhere. Laser sights that emit a continuous or rapidly pulsing beam also offer the advantage of showing you what your barrel is doing in the fractions of a second prior to squeezing the trigger. You can therefore instantly determine whether you are raising the gun too high and lowering it back on target, twisting the gun in your grip as you fire, or making any other minor mistake that could throw your shots off target during those critical microseconds.

One unique way of enhancing practice with laser sights is to use a video camera to record the movement of the laser dot on the target. By playing back the video in slow motion or using a freeze-frame function, you can critically analyze the movement of the dot and the intended point of aim and identify flaws in your technique. This is especially valuable to identify errors such as shoving the pistol or raising it above eye level, then lowering it, before firing.

Although a laser aiming device can be a useful training aid, it suffers from the disadvantage that the laser beam remains constant before, during, and after the trigger squeeze. It is therefore very difficult to determine precisely where a round would impact based on the movement of the laser dot at the instant the hammer drops. Fortunately, Beamhit (8320 Guilford Rd., Columbia, MD 21046) has solved this problem with its ingenious Beamhit training device.

The Beamhit is a battery-powered insert that fits into the bore of a handgun and emits an instantaneous, focused beam of light when the gun is dry fired. Since this light is activated only at the instant the hammer drops, the spot where the light appears on the target is the exact spot where a live round would have impacted.

The Beamhit can be used with virtually any target that will allow the flash of light to be seen; however, it is most effective when used with the Beamhit company's special target system. This photosensitive target actually records the flashes of light and provides an accurate record of where your rounds would have hit. For even greater realism, Beamhit has incorporated this target system into a vest that can be worn by a training partner during tactical training scenarios.

The Beamhit offers a tremendous advantage over both traditional dry-fire practice and dry-fire using laser aiming devices. In recent years it has become extremely popular among law enforcement firearms trainers as an effective yet low-cost training device. In fact, its use has been instrumental in the success of many remedial firearms training programs. For the price of a few boxes of ammo and a couple of batteries, it is one of the best training investments you can make.

Obviously, the best way to learn to shoot a gun well is to practice shooting often. Unfortunately for many, the cost of firing hundreds of rounds through their favorite combat pistol is a bit daunting. However, that doesn't mean you can't have plenty of live fire practice. Since the most important aspects of point shooting remain the same no matter what weapon you're using, you can effectively practice your technique with a .22 caliber pistol or even an air pistol. Ideally you should choose a pistol that closely replicates the feel and balance of your combat gun. If you carry a .45, you can purchase a .22 conversion kit and retain the feel of your carry gun. In recent years, air pistols that closely resemble real firearms have become available, and these make great training aids.

The key to benefiting from practice with these guns is to maintain the same intense mental focus that you would have with a full-bore handgun. The reduced recoil of these weapons could easily cause you to get sloppy and careless in your practice. *Don't allow this to happen.* With a proper mindset, you can cut costs in your training without cutting its effectiveness.

Ultimately, you must devote sufficient time and energy to practicing with full-power combat loads. This gets you used to the recoil of these loads and is the best way to develop the confidence to use a handgun in combat. The key to making the most of this type of training is to make it as realistic as possible. Challenge yourself to shoot under stress, and keep your standards of proficiency within the realm of the practical for an actual gunfight. Many shooters with a genuine interest in developing combat shooting skills subconsciously revert to competition style shooting habits when they practice, striving for tight shot groups and X-ring hits. Avoid this at all costs and instead focus on delivering solid center-of-mass hits quickly and smoothly enough to guarantee yourself a reasonable chance of survival in a real gunfight. To avoid subconsciously trying to focus on your sights and using them to refine your aim off when practicing point shooting, mask them with adhesive tape or consider using a practice handgun with the sights removed or filed off. This will force you to employ true point shooting technique and will speed your progress in mastering this method.

Some useful types of live fire practice include firing at close range targets and multiple targets, firing at angled or partially obscured targets (which represent an opponent behind cover or one facing you indirectly), firing at moving targets, and firing from disadvantaged positions (e.g., off balance, from behind cover, while holding an object in your free hand, etc.).

Because of the increasing use of body armor by criminals, many firearms training programs have begun to include failure-to-stop drills that require head shots as well as center-of-mass hits. Since point shooting can easily enable a shooter to hit an object the size of a dinner plate at typical combat distances, "head-hitting accuracy" for drills of this type is easily achieved. A good practice drill for failure-to-stop scenarios is to fire two rounds at the center of mass, then immediately follow with one or two rounds to the head. A single round to the head is the classic "Mozambique Drill," but a double tap follow-up should also be practiced.

Since most gunfights occur in low light, you should devote considerable amounts of practice to shooting in these light conditions. One excellent method of inducing stress and improving low-light target acquisition skills is the flashlight drill. This drill can be conducted against any berm or embankment and does not require targets.

To do the drill, take up a ready position facing the berm at a typical combat distance. Have a partner with a flashlight stand beside and several paces behind you. At random intervals and without warning, have your partner shine the light at a spot on the berm for a second or so, then turn the light off. As soon as you see the light, you must raise your pistol and fire at the spot before he turns out the light. This drill develops quick reflexes and the ability to pivot and acquire targets rapidly.

Since no target moves quite like a human being (especially one determined to shoot you before you shoot him), the ability to practice man-to-man gunfight scenarios is also an invaluable means of honing your combat skills. Paintball guns shoot gelatin balls filled with brightly colored dyes and are an excellent means of simulating realistic gunfight conditions. Besides enabling you to practice firing on moving human targets, they teach you the proper use of cover in a gunfight. Again, you must strive to maintain an intense mental focus when training in this way to avoid becoming complacent of lackadaisical. If you believe that every paintball that hits you represents a potentially fatal gunshot wound, you will derive much more benefit from this training. You will also find it much easier to simulate levels of stress comparable to a real gunfight. Obviously, any type of training that involves firing projectiles at another person requires proper attention to safety. Paintball training should only be conducted with proper eye and face protection.

A step up from paintball-style gunfight training is the use of a marking round called Simunition. This is basically a self-contained, downloaded cartridge which fires a paint marking round. To use this ammunition in semiautomatic handguns, a special barrel assembly replaces the standard barrel of your handgun, converting it from a locked breech action to a straight blowback action capable of cycling with the light recoil impulse of the Simunition. Thus, unlike the sometimes oversized and unwieldy air-powered paintball guns, Simunition enables you to use your standard sidearm or carry gun to conduct man-to-man gunfight practice. Hits with Simunition also tend to be significantly more painful than paintball hits, further enhancing the realism of the practice and the level of stress of the participants. Again, safety must be emphasized when participating in this type of training. You must follow all the manufacturer's recommendations regarding the use of eye protection and other safety devices to prevent serious injuries when using Simunition.

In a real gunfight, point shooting gives you the greatest chance of surviving the encounter by shooting the other guy before he shoots you. Learn it well and practice it intently, and you'll be ready if the bad guys' muzzle flashes ever point your way.

COMING FULL CIRCLE:
THE REDISCOVERY OF POINT SHOOTING BY LAW ENFORCEMENT

by Steve Barron, Hocking College, Ohio

*P*oint shooting had its roots in the violent armed encounters of early twentieth century Shanghai. Developed out of necessity by W.E. Fairbairn and E.A. Sykes, it provided the members of the Shanghai Municipal Police with an effective method of delivering fire during the gunfights they faced on almost a daily basis.

As a shooting method for the law-enforcement officer, point shooting was undeniably effective. During World War II, it was adapted to the techniques of total war and once again was conclusively proven effective for combat.

Since World War II, however, point shooting as a combat technique has been largely ignored and forgotten. Law-enforcement agencies, in a never-ending effort to modernize their methods and tactics, have adopted newer, more sophisticated shooting methods. Unfortunately, as we have already discussed, the real-world effectiveness of these techniques leaves much to be desired. In short, it wasn't broke, but they fixed it anyway.

Recently, one law-enforcement training facility, confronted by the dismal street performance of officers trained in so-called "modern" shooting methods, decided to reexamine point shooting. They introduced it to groups of new police recruits and conducted objective, side-by-side evaluations of trainees schooled in point shooting and those trained in the modern technique. Not surprisingly, they found that recruits trained in point shooting progressed more quickly and retained their skills much more readily than those trained in the two-handed sighted fire method the school had been using.

The school, Hocking College of Nelsonville, Ohio, is now the pre-mier training facility in point shooting in the United States and is forcing law-enforcement trainers and the entire shooting fraternity to take a hard second look at point shooting. To give you a better idea of how point shooting has come full circle, we offer these comments by Steve Barron, one of the chief instructors of the Hocking College shoot-ing program and proud owner of a changed mind.

The firearms training program at Hocking College has been in operation for more than 25 years. Clyde Beasley has been involved with the program since the beginning; I've been instructing at Hocking since 1978. Clyde and I have been down the road when it comes to law enforcement firearms training. Some of the names we can list on our cre-dentials include the Smith & Wesson Academy, Gunsite, SIGARMS Academy, NRA National Police Firearms Instruc-tor School, and FBI Academy. We are not rookies in the train-ing business, nor are we directed by only one school of think-ing. Another factor that has a positive influence on our approach to training is the fact that we are not "empire build-ing." Our paycheck is the same regardless of what system of shooting we teach or which shooting authorities we quote. We are interested in one thing and one thing only—giving our students the best training we possibly can in the time that we have with them.

From 1978 to around 1992, we taught the so-called "new technique," which incorporated the Weaver stance, a flash sight picture, etc. We were pretty proud of our teachings; after all, that was what the best firearms schools in the nation were teaching. Our basic course for handgun consisted of approx-imately 40 clock hours of training, during which trainees would fire about 600 rounds of ammunition. During the training, the standard exercises included basic sighted fire, smooth draw, low-light shooting, multiple targets, weak hand shooting, and use of cover. In 1990 we added dynamic shooting exercises such as a jungle lane, cotton wad exercise, and scenario training.

After the dynamic exercises were introduced, we soon began noting something dark and ominous, only to be discussed within the confines of our office. We noticed that stance and technique, more often than not, would suffer a total breakdown in any of the shooting exercises that applied some degree of stress on the trainee. This phenomenon occurred with such frequency that we knew something was wrong. Around 1992, we began using Simunition, which only reinforced our earlier observations. It seemed as though our trainees would display an excellent command of the new technique doctrine on the static firing range, but when the stress factor was raised and they were put in a more realistic training environment, technique and stance deteriorated dramatically.

In an effort to correct the problem, we looked to the standard solution of firearms trainers worldwide—reinforce the basics! Well, our trainees became very good and really showed improvement on the static firing range. Take them to the jungle lane, however, and their reinforced basic skills crashed and burned just like all the others.

At Hocking, we have a total of four primary firearms instructors and an equal compliment of defensive tactics instructors on staff. Long ago, we recognized that the firearms and defensive tactics instructors were oftentimes on different sheets of music, so to speak. They often contradicted each other, and the end result was one very confused trainee. To try to eliminate this problem, at Hocking College we require our instructors to cross train and instruct in both areas as often as possible.

For many years, we have adhered to the doctrine of teaching only defensive tactics systems principally based in gross motor movement. We had recognized the fact that the average person cannot perform fine or complex motor skills under stress. As with any competent trainer, we were always looking for a better method or technique.

In the spring of 1995, Clyde Beasley and I were invited to observe a unique defensive tactic training session at the West Virginia State Police Training Academy. The program was presented by American Combatives Inc. from New York. The instructor, Mr. John Kary, presented what turned out to be a very aggressive, no-nonsense unarmed combat course that reminded me very much of my military training.

After the class, I asked Mr. Kary about the origins of the program. He informed me that the course was based on the World War II hand-to-hand combat training as it was taught to the U.S. Marines and U.S. Army special operations personnel. He also advised me that if I wanted more information on the system's origins and development, I should acquire a copy of a book titled *Kill or Get Killed* by Col. Rex Applegate.

A couple of weeks later my copy of *Kill or Get Killed* arrived. Remember, as I opened this book I was thinking defensive tactics. After reading the sections on unarmed combat, I began to browse through the rest of the book. Being both a firearms instructor and enthusiast, the section on combat use of the handgun caught my eye. As I read this section, it was as if someone had just slapped me alongside the head! I was amazed at the fact that I was reading about a system of combat shooting that was based in *gross motor movement*.

I had never really applied the basic rule of keeping all defensive tactics skills rooted in gross motor movement to shooting skills, but it made sense. If we cannot expect a person to duplicate fine or complex motor movements in unarmed combat, then how can we expect it all to change in armed combat?

The following day I made a trip to the firing range. My first act was to tape over the sights on my pistol. Then I put myself through our standard qualification course. The results surprised me. I had done quite well without benefit of sights or using anything that remotely resembled new technique.

For the next couple of days, two other instructors and I worked on developing a more thorough understanding of this shooting system. Each of us had some exposure to point firing during our careers, but none of us had any real comprehensive knowledge of this particular system as it was outlined by Colonel Applegate.

Our next step was to find someone who knew more about it than we did. Our first effort was to try to locate a military vet who trained under Colonel Applegate during the war. While making the usual telephone calls, I was talking with an old friend and associate, Mr. Bert DuVernay of the Smith & Wesson Academy. I asked Bert if he would, in his travels, aid us in our search for someone who had formal training in this system of shooting. Bert's immediate suggestion was, "Why don't you call Rex Applegate?"

I must admit that I was a little intimidated at the idea of just picking up the telephone and giving the Colonel a call. I come from a time in the military when a sergeant just doesn't call a colonel and start asking questions. Well, Bert picked up on this and he took the matter in hand. Within a few days, my telephone rang and Colonel Applegate had my undivided attention.

Colonel Applegate clarified several of the finer points in *Kill or Get Killed* and provided considerable amount of direction in course development. Additionally, he provided some more text materials and videotape. Armed with this material and a few words of wisdom from the Colonel, another instructor and I elected to teach this system to a small class of Police Science students.

In this test group, equal amounts of time and ammunition was spent on point shooting and new technique. At the end of the training cycle, our students were required to qualify on our standard qualification course, once using the point shooting method and again using new technique. Point firing was the clear winner, with a class average of 96.4 percent compared to 81.1 percent.

During this first test run of the point shooting system, we made several observations. First, we were very surprised at how quickly the students were able to develop their shooting skill. Second, we were very pleased with the accuracy level that was achieved. I had always been told that point shooting was impractical because it took too much time and money to develop an acceptable level of skill. We found the opposite to be true. Skills developed quite rapidly and with a minimal expenditure of ammunition.

During this same period, the summer of 1995, we had the opportunity to train 25 in-service law-enforcement officers in point shooting techniques. The results were very encouraging. We administered a pre-test and post-test, and the results showed vast improvements in the shooting skills of about 95 percent of the participants.

In November of 1995, Clyde Beasley and I traveled to Oregon and spent several days with Colonel Applegate. During our visit, the Colonel gave us a full review of the point shooting system, including its history, development, and implementation to military service. We reviewed the training methods that were used during World War II. Many of the finer points of instructional methods were covered, and the technicalities of point shooting were thoroughly addressed. Clyde and I returned to Ohio with a very clear understanding of the point shooting system and the appropriate methods of instruction.

Immediately upon our return to Hocking College, we began teaching point shooting to basic police recruits. These recruits used the technique to qualify on the state's mandatory course of fire. That course addresses such issues as close-quarter shooting, weak hand techniques, multiple targets, low-light shooting, use of cover, and long-range sighted fire.

Eighty-eight percent of this course of fire is inside 25 feet, as it should be. Point shooting simply excelled! Remember, we used to teach the new technique to recruits on this same course, so we have a pretty good standard of comparison.

During the winter of 1996, we taught two more courses. One was a group of basic police recruits, the other a class of in-service law-enforcement officers. These classes again reflected the efficiency of the system. It was of particular interest to note that scores for these groups were quite high, consistently above 95 percent.

In the spring of 1996, Colonel Applegate came to Hocking College and put the finishing touch on our instructional staff. We now have a compliment of four instructors trained in the instinctive point shooting system, and each of them has had the personal attention of the Colonel.

The point shooting program is in full operation now. We are teaching the technique to basic police recruits, police science students, police firearms instructor candidates, and special operations personnel. As of this date, we have provided this training to more than 300 students, and the results have been outstanding.

Our training method is relatively simple. First, we show the student the need for the training. This is accomplished by reviewing the statistical data available on gunfights, including the FBI's "Summary of Officers Killed" and the New York Police Department's records on police shooting incidents. These documents show the student that the clear majority of gunfights take place at extreme close quarters, 20 feet or less, usually in dim light. A quick review of this material, coupled with some additional realities of a gunfight, generally establishes in the student's mind that there is a serious need to learn about close-quarters combat with a handgun.

The next phase of the program consists of giving the student a very detailed explanation of basic safety procedures. This is followed by a lecture on how the excitement and stress

Hocking College of Nelsonville, Ohio, has officially adopted point shooting as its primary handgun shooting method and presently teaches it to all police recruits schooled at their facility. Here a group of recruits practices the combat crouch and instinctive pointing with the index finger during the initial phases of training. By practicing facing a partner, they can note and correct errors in their partner's technique. This is also a safe method of practicing the mechanics of the technique using an actual human target as a visual reference.

of a gunfight affects basic marksmanship skills. This lecture is used to illuminate which shooting system suffers the most from gunfight-induced stress responses.

After the classroom portion of the training is completed, students are taken to the firing range. With all weapons secured in the range house, the now *unarmed* students are lined up in two opposing lines. This is where they will begin learning the basics of stance and pointing. Instruction is given in the forward crouch and the basic low ready position. We drive home the importance of the locked wrist and elbow in

Here the members of a Hocking College class practice the mechanics of point shooting technique on silhouette targets.

combination with the pump handle lift. Students are instructed to use proper technique to raise their hand to eye level and point their index finger at the person opposite them. After a few minutes of this drill, we have the student change the position of his feet (i.e., strong-side-foot forward) and repeat the exercise. This demonstrates to the student that he is not dependent on having his feet in any particular position before he can shoot. This eliminates the tendency for the shooter to do some sort of a change-step to get into the perfect stance prior to shooting. Also covered during this block of instruction is the concept of focusing one's vision on the target.

With pistols in hand, the group is now moved to the firing line, where dry practice on silhouette targets begins. Again, students are schooled in the fundamentals: the forward crouch, locked wrist, locked elbow, pistol alignment with the centerline of the shooter's body, and the vertical lift (or pump handle lift as it's sometimes called). At this time, students are introduced to one of the more critical aspects of this shooting system—the convulsive grip. The importance of this part of the technique cannot be overstated. Having been a firearms instructor for more than 20 years, I've seen the problems that a weak grip can cause when using a semiautomatic pistol. The convulsive grip cures most of them. During this portion of the program, we also address the issue of trigger pull. Students again are given a little dose of reality by telling them that in actual handgun combat, they will most likely fire the pistol by employing the action of the entire hand as opposed to the deliberate trigger press that competition shooters are so fond of.

After the fundamentals are addressed, we move to live fire exercises. At this time, an instructor will conduct a live fire demonstration. *It should be noted that the instructor's pistol has the sights totally removed. This alleviates any question as to whether or not sights were used during the demonstrations.* Students are then placed about 6 feet from the target in a low ready position. The sights of their weapons are covered by tape. On command, they assume a proper firing position and fire one round. This exercise ensures total success, a very important learning tool.

After several shots have been fired, we introduce the student to the concept of firing bursts of two or three rounds. As skill levels begin to improve, the distance is increased. Typically, we see the average student score hits in the upper chest region of the target back to distances of 25 to 30 feet. As distance increases and the score starts to decline, students are instructed to assume a two-handed isosceles firing position but maintain all other point shooting techniques. Generally, this technique will increase the point firing by another 10 to 20 feet. At approximately 50 feet, our students are instructed to switch to the sighted fire position of their choice.

Live fire training at Hocking College begins with point shooting at very close range, then extends the range to the target as the students gain proficiency. Note that the front and rear sights of all the students' handguns have been covered with adhesive tape to ensure that only true point shooting technique is used.

The next phase of the course is the draw and extreme close-quarter shooting technique of the body point. When teaching the body point method, the main focus remains on the forward crouch, convulsive grip, and keeping the forearm parallel to the ground with the elbow firmly locked to the rib cage and the pistol held in alignment wit the centerline of the shooter's body. The latter seems to be the most difficult for students to duplicate under stress. They will develop a tendency to punch the weapon toward the target if not corrected early.

For the drawing sequence, we recommend to use a circular movement of the hand up and under the pistol, secure a firing grip, and lift it from the holster. Care should be taken to

Hocking College students practice point shooting with both the left and the right foot forward to get used to both positions. Here practice is conducted from a crouch with the gun-side foot forward.

insure that a proper firing grip has been established prior to drawing the pistol from the holster. During this phase of instruction, students are encouraged to work on getting the technique correct rather than achieving speed. Speed will come naturally as technique develops. As the students ability at drawing the pistol develops, we move into live fire exercises employing the body point. Emphasis is still placed on the forward crouch, and an aggressive demeanor is encouraged.

After developing a reasonable degree of skill with the body point and draw, the student learns to draw the pistol, pass through the body point position, and move into the point shoulder position. This is a very simple technique. The student draws the pistol, assumes a body point position, then extends the shooting hand until the elbow locks. He is now in a low ready position. With the elbow and wrist locked and a

After becoming familiar with one-handed point shooting, students at Hocking College are introduced to Isosceles-style two-handed point shooting. Again, the technique an individual uses depends upon his or her individual level of skill and the circumstances of the conflict. Note that in this tactical exercise, the nearest shooter uses one-handed point shooting, while the other shooters employ two-handed point shooting technique.

convulsive grip on the pistol, the weapon is raised to eye level and fired center mass of the target.

From this point we move on to such issues as multiple targets, low-light shooting, and target recognition. Once the student is well-grounded in these critical areas, we move on to such issues as barricade and long-range shooting. In these two exercises, we emphasize that when time and distance permit and it is tactically sound, it is proper to take advantage of sights and use traditional marksmanship skill to make a long-range or barricade shot.

As mentioned earlier, we have made several observations

concerning point shooting during the past two years. First, and one of the most surprising for me, is the speed with which novice shooters develop acceptable skill levels. Typically, the basic recruit can be trained adequately in about half the time as was previously used. The remaining time and ammunition can be used to provide tactical training. The next issue is accuracy. This system of shooting provides excellent accuracy in close quarters (inside 30 feet). Third, the speed with which the students are capable of engaging targets is astonishing. Double taps from the low ready in .8 seconds are the average, and it is not unusual to see the first round on target in less than .4 seconds. Remember, this is from a low ready position!

The following are some of the training tools we have found to be quite useful whether teaching basic level students are in-service types:

- A dummy or red-handle pistol equipped with a laser sight is useful for allowing students to access their skill in pointing the weapon without the need to fire live ammunition.
- The old standard of the full-length mirror is a very useful took to allow students to access their stance position (using the dummy pistol).
- A video camera with immediate playback capability is an excellent tool for addressing such issues as improper stance. A Polaroid camera will also work.
- Instructors can make great use of a pistol that has had its sights removed. This eliminates all question as to whether or not the sights were used to make the shot when demonstrating point shooting.
- Subcaliber training weapons are somewhat useful in the early stages of training.
- Simunition or Code Eagle ammunition is an excellent training tool. *Remember safety first* when using this ammunition in training.

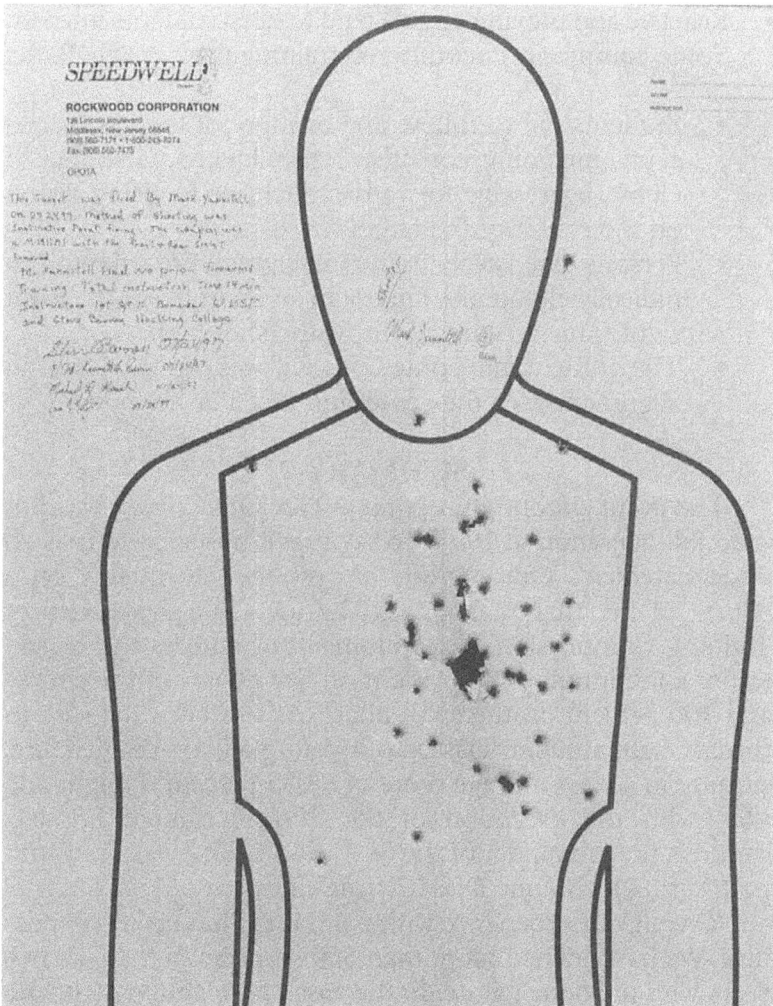

This target was shot by Mark Yannitall, videographer for Hocking College. After taking video of numerous shooters participating in point shooting training, Yannitall, who had never fired a handgun before, asked if he could give it a try. After only 14 minutes of instruction, he shot this target with an M1911A1 that had its sights removed. The proof is in the shooting!

- Reactive and moving targets tend to keep students interest. Some commonly encountered training problems include:

  - Students fail to follow one or more of the basic concepts, including convulsive grip, locked wrist, locked elbow, aggressive forward crouch, or focusing vision on the center of the target.
  - Persons extensively trained in sighted fire will tend to radically elevate the muzzle in an effort to see the front sight. This action will throw the shot high.
  - The student will sometimes fail to keep the pistol in alignment with the centerline of his or her body.

SUMMARY
The point shooting program at Hocking College was put in to full operation in 1995. To say that it was successful is an understatement. Our recruits are required to qualify on a course of fire that is mandated by the Ohio Peace Officer Training Commission. The commission requires the recruit to fire a minimum of 80 percent on five phases of the course and 100 percent on the sixth phase. As has been the case in the last eight academy classes, our young officers did just fine, turning in a class average score of 98.33 percent. I might add that, with the exception of the 50-foot sighted fire (six rounds), the remainder of the 56-round course was fired with point firing technique *with the sights taped over*.

Overall, our experience with point firing has been very positive. We have trained more than 500 shooters in the past two years with only one failure. In the case of the young man who failed, it should be noted that he passed all close-quarter exercises—it was the 50-foot sighted fire that he could not grasp.

Here at Hocking we also train police firearms instructors. During the past two years, we have trained around 50 instructors. These officers are exposed to various methods of shooting, including Weaver, Isosceles, and point shooting. They are

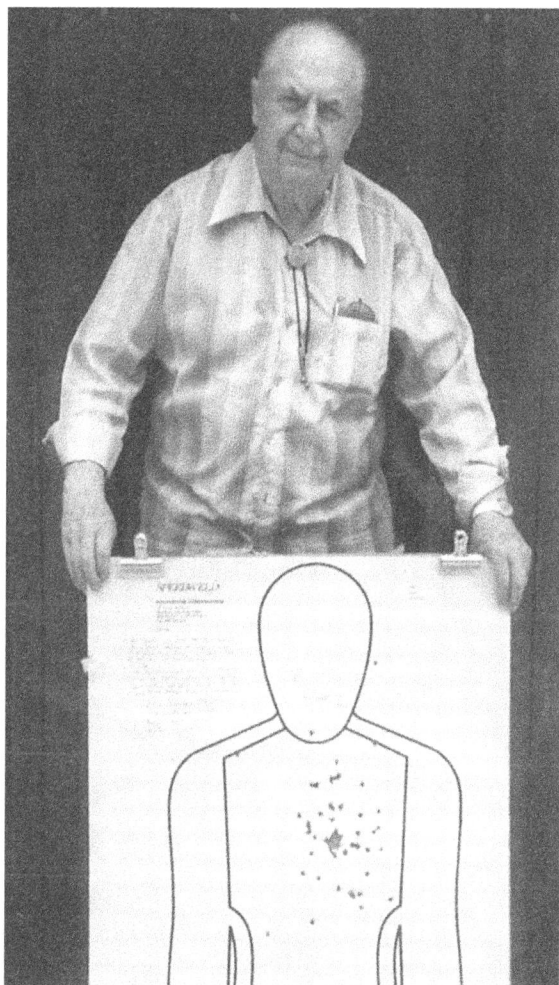

Col. Rex Applegate, 1997.

astonished that hits can be achieved with point firing techniques at distances well beyond the 20-foot line. Of all the
instructor candidates completing these programs, only one
would not accept that point firing techniques worked, even
though he fired a score in the upper 90s with his sights covered by tape!

Point shooting was developed to win gunfights, not competitive games. It was developed for average people, not gun enthusiasts. Most importantly, it is combat proven. Simply put, *it works!*

Until some one can show us something better, this system will be in place at our institution for a long time to come.

• • • • •

For more information about point shooting classes offered by Hocking College, contact:

Hocking College
Department of Public Safety Services
Police Science Technology
3301 Hocking Parkway
Nelsonville, Ohio 45764

SUGGESTIONS FOR FURTHER STUDY

The following book and video titles contain additional information concerning the technique of point shooting, its history, and its combat application.

Applegate, Rex. *Kill or Get Killed*. Boulder: Paladin Press, 1976. This is *the* classic text on all aspects of close combat with and without weapons. Originally written during World War II and adopted as an official training manual by the U.S. Marine Corps, it includes the original instructional material on Col. Applegate's method of point shooting as he taught it to OSS and U.S. military intelligence operatives.

Applegate, Rex. *Point Shooting: Battle-Proven Methods of Combat Handgunning*. Paladin Press, 1995. This video presents the declassified World War II training film Film Bulletin 152, the only official U.S. government training film ever produced on point shooting. The film presents instruction in point shooting technique as it was taught during World War II and is accompanied by an historical introduction by Colonel Applegate.

Applegate, Rex. *Shooting for Keeps: Point Shooting for Close-Quarter Combat.* Paladin Press, 1996. This is the most complete and detailed instructional video ever produced concerning point shooting. This video closely parallels the text of this book and includes live-fire demonstrations of the techniques detailed herein.

Cassidy, William L. *Quick or Dead.* Boulder: Paladin Press, 1993. This well-researched book presents a complete history of the evolution of point shooting and the men and events that influenced its development.

Fairbairn, William E. and Sykes, Eric A. *Shooting to Live.* Boulder: Paladin Press, 1987. This is a reprint of Fairbairn and Sykes' original work *Shooting to Live with the One-Hand Gun*, published in 1942 and detailing the point shooting method they developed during their service with the Shanghai Municipal Police.

Siddle, Bruce K. *Sharpening the Warrior's Edge: The Psychology and Science of Training.* Millstadt, IL: PPCT Research Publications, 1995. Bruce Siddle is an accomplished law enforcement trainer and researcher who has pioneered the application of scientific research and testing methods to the realm of close combat. This book presents his insights into the physiological and psychological phenomena that occur during life-threatening encounters and provides scientific validation of the value of point shooting in high-stress situations.

An Important Message from Author Michael Janich

Bullseyes Don't Shoot Back was originally published by Paladin Press in 1998. For me, co-authoring a book with the legendary Col. Rex Applegate was a tremendous honor and privilege. Although Col. Applegate passed away shortly after the book was released, *Bullseyes* went on to establish itself as a classic reference text on the Colonel's approach to point shooting in its most refined form.

Following Col. Applegate's death, my personal approach to defensive handgun skills continued to evolve. While I will always remain deeply influenced by Col. Applegate and his teachings, I was also inspired by other iconic instructors like Jim Cirillo and Kelly McCann, with whom I also worked very closely during my years at Paladin Press.

In 2017, Paladin Press closed its doors and the rights to this book reverted to me and Col. Applegate's widow, Carole. Initially, I thought it best to simply take pride in the fact that the book had remained in print for nearly 20 years and had helped thousands of readers learn practical, usable defensive shooting skills. However, after considerable reflection—and with Mrs. Applegate's generous help and support—I decided to bring this book back into print to honor Col. Applegate's memory, to continue the legacy of his teaching, and because of its continued importance as a historical reference.

If you enjoyed this book and are serious about learning other practical, reliable self-defense skills, I encourage you to visit www.martialbladeconcepts.com and https://martialbladeconcepts.tv/shop/. These web sites provide more information about my personal system of defensive shooting and my instructional videos, including my Paladin Press "classic" videos, my more recent videos produced by Stay Safe Media, and my Martial Blade Concepts Distance Learning Program.

Thank you for purchasing this book. I hope the information we shared will help you further your personal training goals and keep you and your loved ones safer. I also hope to have the opportunity to train with you sometime soon.

Stay safe,

Michael Janich

www.ingramcontent.com/pod-product-compliance
Lightning Source LLC
Chambersburg PA
CBHW071208200326
41519CB00018B/5426